COMPLIANCE OR DEFIANCE?: ASSESSING THE IMPLEMENTATION OF POLICY
PRESCRIPTIONS FOR COMMERCIALIZATION BY WATER OPERATORS

Mireia Tutusaus

COMPLIANCE OR DEFIANCE?: ASSESSING THE IMPLEMENTATION OF POLICY
PRESCRIPTIONS FOR COMMERCIALIZATION BY WATER OPERATORS

ACADEMISCH PROEFSCHRIFT

ter verkrijging van de graad van doctor
aan de Universiteit van Amsterdam
op gezag van de Rector Magnificus
Prof.dr.ir.K.I.J. Maex
ten overstaan van een door het College voor Promoties ingestelde commissie,
in het openbaar te verdedigen in de Agnietenkapel
op 10 december 2019, te 14:00 uur

door Mireia Tutusaus Luque
geboren te Barcelona

Promotiecommissie

Promotor	Prof. dr. M. Z. Zwarteveen	Universiteit van Amsterdam
Copromotor:	Dr. K.H. Schwartz	IHE Delft Institute for Water Education
Overige leden:	Prof. dr. I.S.A. Baud	Universiteit van Amsterdam
	Prof. dr. J. Gupta	Universiteit van Amsterdam
	Prof. dr. R.A. Boelens	Universiteit van Amsterdam
	Prof. dr. R. Hope	University of Oxford
	Dr. K. Furlong	Université de Montréal
	Dr. M.A.C. Schouten	VEI Dutch Water Operators

Faculteit der Maatschappij- en Gedragswetenschappen

CRC Press/Balkema is an imprint of the Taylor & Francis Group, an informa business

Published by:
CRC Press/Balkema
Schipholweg 107C, 2316 XC, Leiden, the Netherlands
Pub.NL@taylorandfrancis.com
www.crcpress.com – www.taylorandfrancis.com
ISBN 978-0-367-89511-2 (Taylor & Francis Group

Naleven of tarten? Een analyse van hoe beleidsvoorschriften voor de commercialisering van waterleveranciers worden uitgevoerd

De 'mondiale watergemeenschap' is opvallend unaniem in het denken over de wijze waarop universele toegang tot drinkwatervoorziening bereikt kan worden. De diagnose waarop de beleidsconsensus is gebaseerd is dat waterleveranciers niet genoeg fondsen aantrekken om de ontwikkeling en het onderhoud van waterinfrastructuur te financieren. Dit is waarom het nastreven van financiële duurzaamheid centraal staat bij beleidspogingen om de watervoorziening te verbeteren. Financiële duurzaamheid stoelt op twee met elkaar verweven principes: autonomie van de waterleverancier en het vermogen om te werken op basis van volledige kostendekking. Samen vormen deze principes de belangrijkste pijlers van wat vaak kortweg aangeduid wordt als commercialisering.

Commercialisering gaat ervan uit dat het vermogen om op basis van volledig kostendekking te opereren een voorwaarde is om universele waterlevering te bewerkstelligen. De gedachte is dus dat de financiële doelstelling van volledige kostendekking verbonden kan en moet zijn met de sociale doelstelling van universele dienstverlening. Hoewel verschillende auteurs en empirische studies deze veronderstelling in twijfel trekken, blijft de overtuiging in de drinkwatersector dat beide doelen complementair en haalbaar onverminderd sterk.

Het is aan de werknemers van waterbedrijven om commercialisering – en de principes van autonomie en kostendekking - in hun dagelijkse werkzaamheden te realiseren. Het blijkt dat in de dagelijkse praktijk, commercialisering kan conflicteren met het mandaat om iedereen van water te voorzien. Ook kan de sociaal-politieke, economische en bio-fysieke context waarin waterleveranciers opereren, de implementatie van commercialisering moeilijke maken. Dit onderzoek laat zien hoe in hun pogingen de beleidsprincipes na te leven, drinkwaterbedrijven vaak gedwongen zijn deze aan te passen, opnieuw te interpreteren of zelfs helemaal te negeren. Het onderzoek is geïnspireerd op inzichten uit de literatuur over beleids- implementatie en -vertaling. Deze literatuur stelt dat beleidsmodellen moeten worden aangepast om succesvol te kunnen worden overgenomen of geïmplementeerd. In dit onderzoek beschouw ik een serieuze beschrijving van deze veranderingen als een manier om zichtbaar te maken wat commercialisering in de praktijk van drinkwaterbedrijven betekent.

Dit onderzoek kijkt naar drie casussen die de drie dominante modellen van waterdienstverlening vertegenwoordigen: het commerciële publieke waterbedrijf, de uitbesteding van activiteiten aan de private sector en de levering van water via kleinschalige gemeenschaps organisaties. In Mozambique beschrijf en analyseer ik de casus van AIAS en de uitbesteding van activiteiten aan kleinschalige particuliere exploitanten. In Oeganda wordt de casus van de National Water and Sewerage Corporation, die als een commercieel publiek nutsbedrijf werkt, onderzocht en in Indonesië analyseer ik de uitbreiding van waterdiensten via kleine gemeenschapsorganisaties die zijn opgericht om als commerciële exploitanten te opereren. Bij de selectie van deze casussen heb ik er doelbewust voor gekozen om me te concentreren op waterlevering in kleine steden of (semi) rurale gebieden, omdat juist hier commercialisering bijzonder uitdagend. Kleine steden vertonen drie specifieke kenmerken die het aanbieden van waterdiensten op commerciële basis moeilijk maken. Ten eerste is de bevolkingsdichtheid in kleine steden is over het algemeen lager dan in grote stedelijke gebieden, waardoor het voor exploitanten moeilijk is om te profiteren van de voordelen van schaalvoordelen bij de ontwikkeling van infrastructuur. Door water te leveren via een leidingnetwerk aan een klein consumentenbestand dat ook nog eens verspreid is over een relatief groot leefgebied, zijn deze systemen relatief duur in gebruik. Ook is het besteedbare inkomen van consumenten in deze gebieden over het algemeen lager dan in de grote steden. Een tweede uitdaging is de technische en financiële capaciteit die lokaal beschikbaar is om watersystemen te ontwikkelen en te exploiteren. Ten derde vereist de dynamische en relatief snelle groei van kleine steden vaak meer flexibele water systemen, in plaats van de traditioneel rigide zuiveringsinstallaties en distributienetwerken. Al met al maken deze factoren samen het moeilijke om volledige kostendekking te realiseren in kleine steden.

De casussen die in dit onderzoek worden geanalyseerd tonen aan dat waterleveranciers de principes van commercialisering op verschillende manieren opnieuw hebben geïnterpreteerd om ze in hun context van kleine steden te kunnen implementeren. Deze aanpassingen betwisten zowel het idee dat deze waterleveranciers als autonome entiteiten (kunnen) opereren en dat ze in staat zijn dit te doen terwijl ze de kosten volledig terugverdienen. Mijn onderzoek laat zien hoe waterleveranciers juist actief proberen onderdeel te zijn en blijven van politieke en beleidsdomeinen. Dit is omdat nauwe samenwerking met politici en overheidsinstanties deze drinkwaterbedrijven toegang geeft tot extra financiering die ze gebruiken om hun (niet-kostendekkende) activiteiten te subsidiëren.

Ondanks deze aanpassingen, beschouwt de internationale beleidswereld commercialisering niet als on- of maar deels haalbaar. Integendeel, het geloof in het model van commercialisering blijft overeind en wordt zelfs gevierd - niet alleen door donoren, maar door overheidsinstanties, en de waterbedrijven zelf. Ik verklaar dit fenomeen door te stellen dat het commercialiseringsmodel dat wordt gepromoot door donoren en overheidsinstanties, en het model dat wordt geoperationaliseerd door waterleveranciers, verschillende politieke domeinen en verschillende doelen dienen. In de mondiale en nationale beleidsdomeinen vertegenwoordigt het commercialiseringsmodel een overeengekomen aanpak, een manier voor donoren, overheden en waterbedrijven om hun belangen samen te brengen. Deze actoren zijn afhankelijk van elkaar en ze zijn allemaal gebaat bij de weergave van commercialisering als een succesvol beleid. Door te laten zien en te herhalen dat commercialisering werkt kunnen donoren beweren dat ze succesvolle projecten ondersteunen, terwijl nationale regeringen kunnen beweren dat ze universele dienstverlening nastreven. Enigszins paradoxaal, ook voor waterbedrijven is het nuttig om zich actief te presenteren als conformerend aan het beleidsmodel, omdat het hen in staat stelt toegang te krijgen tot aanvullende financiering. In dit beleidsdomein is het minder belangrijk of in de dagelijkse dienstverlening, waterleveranciers in kleine steden zich ook echt aan het commercialiseringideaal kunnen houden. Hoewel waterleveranciers het model willens en wetens aanpassen en wijzigen om hun consumenten van water te kunnen voorzien, is het niet in hun belang om het commercialiseringsmodel openlijk aan te vechten. Het gevolg is dat er een duidelijke verschil is tussen het commercialiseringsmodel zoals dat in mondiale en nationale beleidsdomeinen wordt gebezigd en wat drinkwaterbedrijven doen in hun dagelijkse dienstverlening.

Dit proefschrift betoogt dat de dagelijkse praktijk en werkelijkheid van drinkwaterbedrijven veel serieuzer moet worden genomen in het nadenken over hoe drinkwaterveroorziening verbeterd kan worden. Ik laat zien dat wat drinkwaterbedrijven doen niet uitgelegd kan worden in de taal van het commercialiseringvertoog, maar een nieuw vocabulaire vereist. Momenteel wordt de commercialisering van waterleveranciers gezien als een doel, een belangrijke prestatie-indicator voor een waterbedrijf, in plaats van als een middel om alle consumenten van voldoende en veilig drinkwater te voorzien. Veel van de beleidsgeoriënteerde literatuur lijkt te suggereren dat het voor een niet-commerciële wateroperator bijna onmogelijk is om efficiënt of effectief te werken. Wat de in dit proefschrift geanalyseerde praktijken echter laten zien is dat er meerdere manieren zijn om commercialisering te belichamen en uit te voeren. Door de praktijk van waterleveranciers serieus te nemen, kan ruimte worden gecreëerd voor het nadenken over en ontwerpen van adequatere en eerlijkere regelingen voor het verlenen van waterdiensten. In deze ruimte moeten overheden en donoren verder gaan dan normatief-technische assistentieprogramma's die erop gericht zijn exploitanten die niet voldoende commercieel zijn te 'corrigeren'. Financiële duurzaamheid, operationele efficiëntie of de prijs van water blijven belangrijk, maar er moet veel meer ruimte zijn om subsidies openlijk te bespreken als een vaak noodzakelijke bron van inkomsten, of om te erkennen dat politieke inmenging plaatsvindt, zonder dat dat noodzakelijkerwijs schadelijk hoeft te zijn voor de kwaliteit van geleverde diensten.

Compliance or Defiance? Assessing the implementation of policy prescriptions for commercialization by water operators

The "global water community" is strikingly unanimous in stating what is required to achieve universal access to drinking water services. Anchored in the belief that the lack of progress towards achieving universal coverage lies in the inability of water providers attract and retain funds for the development and operation of water infrastructure, these actors have made financial sustainability a key objective to be achieved by any water provider. For these actors financial sustainability mainly revolves around two intertwined principles: autonomy of the water services provider and the ability to operate on the basis of full cost recovery. These two principles form the main pillars of commercialization.

In promoting commercialization, the ability to operate autonomously and on the basis of full cost recovery is argued to be a condition for a water provider to ensure universal service coverage. As such, the financial objective of achieving full cost recovery is portrayed as being interconnected with the social objective of universal service coverage. Such a view point suggests that water providers should be able to fulfil a mixed mandate of achieving universal service coverage, whilst doing so in a financially sustainable manner. Although various authors and empirical studies question the assumption that water providers are able to fully achieve both objectives, the idea of the mixed mandate has become dominant in the water services sector.

Whereas the global water community may have reached consensus on the need for water providers to operate on the basis of commercial principles, staff of water utilities are faced with the challenge of implementing these principles in their everyday work. In the everyday domain, these principles appear to directly conflict with the mandate of water operators to provide water services to all. Moreover, the socio-political, economic and bio-physical context in which these water operate may be ill-suited to implement commercialization. In pursuing commercialization these operators adapt, reinterpret, modify, deflect, alter or betray the original principles of commercialization during implementation. This research takes inspiration from the rich literature on policy implementation and policy translation, which argues that policy models need to be transformed and modified if they are to be successfully adopted or implemented. In this research I analyze the alterations visible in the daily implementation of commercial models of water provisioning and, in doing so, present a better understanding of how water operators implement policy prescriptions of commercialization in practice.

This research is based on three cases that represent the three dominant models of water service provision. These three models are the commercial public water utility, the delegation of operations to the private sector, and provision of water through community-based organizations. In Mozambique I document the case of AIAS and the delegation of operations to small-scale private operators. In Uganda the case of the National Water and Sewerage Corporation, which operates as a commercial public utility, is examined and in Indonesia I analyze the expansion of water services through community-based organizations that have been established to operate as commercial operators. In selecting these cases I have purposefully opted to focus on water services in small towns or (semi)rural areas. The reason for selecting small towns is that the requirements of commercialization are particularly challenging in the context of small towns. Small towns present three specific dilemmas in terms of the provision of water services. The distribution of people in small towns is generally less dense than in large urban areas which makes it difficult for operators to reap economies of density and economies of scale in infrastructure development. Providing water through piped networks to a small consumer-base that is also scattered over a relatively large area makes these systems relatively expensive to operate. Also, the disposable income of consumers in these areas is, generally, lower than in typically urban areas. A second challenge is presented by the technical and financial capacity available at the local level to develop and operate water systems. Third, the dynamic and relatively fast growth of small towns often requires the implementation of more flexible structures rather than established designs of treatment plants and distribution networks. These factors make operating on the basis of full cost recovery a serious challenge.

The cases presented in this research show that water operators have, in different ways, reinterpreted the principles of commercialization in order to be able to implement them in their small town contexts. These adaptations challenge both the notion that these water provider operate as autonomous entities and that they are able to do so whilst fully recovering costs. Instead it appears that it is the ability of the water providers to actively engage with the political and policy domains that allows them to access additional funding, which can be used to subsidize their operations.

Despite the adaptations to commercialization, these modifications are not viewed as compromising or challenging the model of commercialization. Rather, the model of commercialization remains dominant and even celebrated by government agencies, donors and even the water operators themselves. In this research I explain this phenomenon by arguing that the model of commercialization promoted by donors and government agencies and the model as operationalized by water providers cater to different political domains and serve different purposes. In the global and national policy domains, the model of commercialization represents a negotiated approach that serves the interests of donors, governments and water utilities. These actors are dependent on each other and a successful portrayal of commercialization serves their interests. Donors can claim that they support successful projects, national governments can claim that they are pursuing universal service coverage and, somewhat paradoxically, water operators that manage to present themselves as commercial water providers gain access to additional funding. Whether, in the everyday domain of service provision, water providers in small towns are able to adhere to this ideal is less important. Although, water providers need to adapt and modify the model in order to provide services to their consumers, it is against their interest to openly challenge the model of commercialization. What ensues is a clear disconnect between the discursive model of commercialization in global and national policy domains and an operationalized model in the everyday domain.

Based on the analysis of the adaptations and (re)interpretations of the implemented model of commercialization in the different cases, this thesis argues that a new way of speaking about commercialization should be developed. This new vocabulary should start from the everyday practices of operators. Currently the commercialization of water services is seen as an end, an important marker of performance for a water utility, rather than as a means to provide adequate and safe drinking water to all consumers. Much of the policy-oriented literature seems to suggest that it is almost impossible for a non-commercial water operator to be able to operate efficiently or effectively. However, what the practices analyzed in this thesis reveal is that there are multiple ways of embodying and doing commercialization. When actual practices of water operators are seriously taken into consideration from a place of empathy (if not sympathy), space for more honest arrangements for providing water service may be created. In this space the governments and donor community may go beyond technical assistance programs that try to 'correct' operators that are not sufficiently commercial. Financial sustainability, operational efficiency or the pricing of water can continue to be acknowledged as important, but there should be much more room to openly discuss subsidies as a very much needed source of support, or to acknowledge that political interference happens, without necessarily being detrimental to the provision of water services.

Financial Support

Data collection for this research was initiated in June 2015 and partially funded by an independent small grant (30.000 EUR) from the Ministry of Infrastructure and Environment (currently Water and Infrastructure) of the Netherlands. This research has been continued from 2016 until 2018 under the framework of the SMALL project (Water Supply and Sanitation in Small Towns) funded under the Programmatic Cooperation between the Ministry of Foreign Affairs of the Netherlands and IHE Delft Institute for Water Education (DUPC2). IHE Delft granted Mireia Tutusaus 400 hours for research in 2016 to dedicate to the development of this.

Acknowledgments

What a trip! I doubted myself until a few days before writing this lines that I was ever going to see the end of this. Of course, I will honor all clichés. I would not have been able to do this alone. And I was very lucky to be around so many very stubborn and perseverant, encouraging and never to be exhausted wonderful people.

First of all, an enormous thank you goes to my two supervisors. Margreet, I admire many things about you. Above all, your intellect and sharpness mixed with deep empathy. I want to think that some of it seeped into my brain and pores through the many, many incessant rounds of comments you provided and the formal and less formal talks we had. You lifted me up. Klaas, we disagreed on many things along the way, but I thoroughly enjoyed each and every step. I am grateful for your perseverance, your creativity and of course for the sharing of 'the classics' with me. You told me I could, and at some point I really believed you. Thank you. We did a lot of 'abstracting ideas' these last years, and I think what we often particularly did best was abstracting ourselves from the rest of the world with our magnificent ideas. Let's continue doing some more of that in the years to come, wherever we are.

If it was not because in the early years of this research I met Giuliana, this thesis would most likely not have been written. You insisted (you are not at all stubborn!) in submitting projects related to a topic in which we could work together. We got them! And I could finally spend time reading and writing which was very much needed. This thesis is then also here because of you. I am grateful you pushed me and you took on the often frustrating project management of SMALL, all while I was writing away…In the meantime you also became a very good friend and a family figure to Ona and Lis, which to me, is not a bit less important than the rest. Thank you. This one also goes to Marcus.

In my years at IHE I have very much enjoyed my interaction with students and I owe them a big chunk of my professional and personal development. A very special thanks to Riski and Maxi because without your support, trust and generosity I could not have completed this thesis. I hope I made justice to your work, and I will always acknowledge your contribution.

A thank you to Maria and Rhodante, first for taking me to Mozambique to then unknown world to me of 'doing research' for the first time and for always, from close or far away, inspiring me to see things differently. You have seen me cry and you have seen me laugh and I remember either of these moments dearly. You are both something special, and I keep you very close to my heart. These words do not do justice to what you have meant to me in submitting this PhD thesis and in doing the work I do, and how I do it.

To my colleagues at IHE, specially the Water Governance chairgroup, thank you. You took me in as one of you even though I was for most part an outsider. It never felt like that and it meant a lot to me. You have helped me navigate and survive the IHE machinery. Thanks for the laughs. Jos, I am so very happy you were there too. A special thanks to Andres, for never shying away from sharing your dramatic and wise life-lessons. Brindemos! Also, thanks to Janez who battled together with Klaas to get me a place at IHE. Susan, you have always been a role model to me for the work at IHE, but much more for the 'work' outside of IHE, and the combination of the two!. Thank you.

I am grateful to my parents who despite never really grasping the idea of what I was doing (my bad for not being more eloquent), they never for one second spared in support. Your unconditional support cannot be described in words and it is also through finalizing this PhD thesis, like in all the other milestones in my life, that I realize how lucky I am to have you as my parents. This applies to my brother as well. If I have any analytical skills and any capacity to build arguments today is because I have been closely 'trained' and very much challenged by my dear and very, very loving brother… Gràcies. To my uncles and aunts in Spain for keeping my parents sane in the process, then, now and always. To my younger cousins, thank you, because I always thought if I could finish this, I could tell you, you could do and be whatever it is you want to do and be. No limits.

Before this process started, during, and I hope after, it has always been essential to me go to Luca and Sara for anything else related to life. But most importantly, you reminded me along the way where I come from and what is possible (and also what was not possible!). I am grateful I can consider you my friends.

To my very, very beloved family in the Netherlands, thank you for your immense support in helping me push this through and for always taking care of me, Jaap, Ona and Lis when I could not.

Jaap, Ona and Lis, I did this *with* you and I would not want to have done it any other way. Let's continue having fun!

Gràcies! Thank you! Dank jullie wel!

Table of Contents

Acronyms

ADB	Asian Development Bank
AIAS	Administracao de Infra-structura de Agua e Sanemento
ATWATSAN	Alternative Technologies for Water and Sanitation Project
CRA	Conselho Regulador de Agua
CBO	Community-Based Organization
DMF	Delegated Management Framework
DNAAS	National Water Directorate
DPOHRH	Provincial Offices of the Ministry of Public Works and Housing
DUPC2	DGIS UNESCO-IHE Programatic Cooperation II
FIPAG	Fundo Investimento e Patrimonio do Abastecimento de Agua
GOU	Government of Uganda
HIPPAM	Himpunan Penduduk Pemakai Air Minum
IHE	Institute for Water Education
ISDP	Infrastructure Service Delivery Plan
KRIP	Kampala Revenue Improvement Programme
MCC	Millenium Challenge Corporation
MDG	Millenium Development Goals
NWSC	National Water and Sewerage Corporation
ODA	Official Development Assistance
OECD	Organization for Economic Cooperation and Development
PACE	Perfromance Autonomy Creativity Enhancement Contracts
PDAMs	Perusuhaan Daerah Air Minum
PPP	Public-Private Partnerships
PSP	Public Stand Pipes
SDG	Sustainable Development Goals
SMALL	Water and Sanitation in Small Towns Project
VEI	Vitens Evides International
WHO	World Health Organization
WSP	Water and Sanitation Program of the World Bank
WSSP	Water Supply Stabilization Plans

List of Tables

List of Figures

Chapter 1: Introduction

For many actors in the water services sector financial sustainability has become a defining mantra in the provision of water services. Government agencies and (international) donors that claim to support the development of services around the world see financial sustainability of water service operators as the cornerstone to achieving universal service coverage. This endorsement of achieving financial sustainability essentially revolves around three arguments. The first argument establishes a direct relationship between revenue generation and the quality of service provided by the water operator. It is only through the 'right' pricing of water that providers can provide adequate and safe water services. Utilities that do not have the capacity to generate their own revenue are portrayed as being trapped in a "low-level equilibrium" of performance (Spiller and Savedoff, 1999). Utilities that are stuck in this equilibrium do not generate sufficient revenues from operations to invest in service improvement. This low level of service leads to unsatisfied customers that, in turn, are not willing to pay for services. The unwillingness of the customers to pay perpetuates the inability of the utility to generate sufficient revenues and 'escape' low levels of performance. A second, related, argument relates to the limitations imposed on the utility in deciding about financial matters. The core of this argument is that "ineffective and misdirected policies" governing the sector (Baietti et al., 2006:1) limit the ability of the water provider to make the necessary decisions to escape the low performance equilibrium. If utilities were sufficiently autonomous to determine cost-recovering tariffs and pursue efficient water operations, they would be able to break the dependency on external subsidies and donor programs, allowing them to assert more control on where and how investments are made. A third argument, finds that water operators should be encouraged to rethink and diversify the financial streams and sources of money supporting the development of water services. Given the expected decrease in donor assistance for the water services sector in developing countries, the utilities should give more importance to commercial and blended finance (UN-Water, 2018). To achieve this, water utilities need to be able to show that their operations are interesting investment projects. In order to mobilize commercial finance for the water services sector "reforms are required to promote efficiency gains, cost reduction and cost recovery in the water services sector", as well as "[improve] the balance of tariffs and taxes as sources of finance"' (OECD, 2018: 11). Public funds, which currently still make up a large portion of investment funds, would then no longer be used to supplement shortcomings of the water utilities as a result of inefficient operations or even fund the development of water works directly, but rather be used to provide the guarantees for private investors safeguarding the risks-return associated with water investments (UN-Water, 2018).

As the brief description above illustrates, the emphasis on financial sustainability is closely linked to the principles of autonomy and full cost recovery. Together these two principles constitute the pillars of "the commercialization of water utilities" (Kitonsa and Schwartz, 2012). Over the past decades, these principles have become engrained in the 'global water community'[1] and, as a result, also in the policy prescriptions emanating from this community. In these policy prescriptions commercialization has seemingly become undisputed and absolute (Rusca and Schwartz, 2018).

1. Problem statement: Financial sustainability and universal services

Most water utilities are not only subject to policies directing them to provide sustainable water services. These providers are also the main vehicle of national governments to achieve universal service coverage (Baietti et al., 2006). In policy-oriented literature, financial sustainability and the extension of coverage are frequently presented as mutually dependent (Rusca and Schwartz, 2018). Only by achieving financial sustainability can a water utility achieve universal service coverage. The water provider´s "inability to set cost recovering tariffs, often under the guise of achieving social objectives, constrains water utilities' capacity to expand services to low-income areas" (Gerlach and Franceys, 2008 as cited in Rusca and Schwartz, 2018:102). The compatibility of financial cost-recovery and universal service coverage is also reflected in policy documents. Illustrative of this line of reasoning is the Kampala Statement that was drafted at an international conference in Kampala (Uganda) organized by the Water Utility Partnership[2] in 2001. The Kampala Statement first concludes

---

[1] I use the term 'global water community' to refer to the actors that engage in international policy domains. These actors, who frequently meet in events such as the Stockholm Water Week or the Biannual IWA Conference, may include international donors and agencies, national government ministries and agencies, knowledge institutes, and large water utilities.
[2] The Water Utility Partnership is an African regional capacity building network focusing on the improvement of water services to urban areas. This network is formed by, among others, the African Water Association, the Regional Center for Low Cost Water and Sanitation in Burkina Faso, the Training, Research and Networking for

that the "needs [of the poor] are often overlooked in the design of various reform programmes". Subsequently, however, the Statement explains that "improved cost-recovery, to ensure sustainability and improve service" is "one of the cornerstones of water and sanitation reform" (Kampala Statement 2001:3). Most development projects in the water services sector aimed at improving water coverage matter-of-factly combine the extension of services (to low-income households) with financial sustainability (Kessides, 2004; Rothstein and Galardi, 2007).

Critical scholars have, however, questioned the compatibility of recovering costs and expanding water services to reach universal coverage. Furlong (2015:206), referring to this 'mixed mandate' of public water utilities, questions the possibility to achieve both social and commercial objectives simultaneously. Others have likewise identified the combination of social and commercial objectives as an oxymoron, leading to a "confused identity" of water utilities (Loftus, 2004:73) or "the schizophrenia of public enterprises" (van Rooyen and Hall, 2007: 60). Some authors refer to the dichotomous binaries of rights/commodity and citizen/customer to qualify these two mandates as intrinsically opposite, suggesting their incompatibility (Bakker, 2007:433). Different empirical studies suggest that the expectation to operate as a financially autonomous utility provides "adverse incentives to the extension of networks to poorer areas, as both private operators and corporatized utilities can perceive them as unprofitable markets" (Marson and Savin, 2015: 27). Berg and Mugisha (2010:592) similarly argue that "the utility does not have strong incentives to build new connections [...] without external funding". Rather "utilities [in Africa] tend to target 'high priority' areas for expansion where immediate financial returns are more promising" (Castro and Morel, 2008:291; see also WSP 2009). This more critical literature, thus, suggests that instead of being mutually reinforcing, combining universal service coverage with financial sustainability may be challenging to water operators. In other words, simultaneously being the main vehicle for expanding water services and operating as a financially autonomous entity may be difficult goals to reconcile (Furlong, 2010).

2. Research justification

Even if the two objectives appear difficult to reconcile, water operators are held accountable for their progress towards achieving both aims. As such, water operators have to, to the best of

Development in Ghana and supported by the World Bank (http://web.mit.edu/urbanupgrading/waterandsanitation/introduction/wup.html - Accessed 03 August 2019).

their capabilities, find ways of dealing with this mixed mandate. The question of how water operators try to balance financial and social objectives was a larger theme of work covered by the Water Services Management Group – Water Governance at IHE Delft. In our lectures we frequently review similar ideas as those discussed in this and subsequent chapters of this thesis: the challenges of reconciling universal access to water services with financial sustainability of water operators. We discuss how principles of commercialization and financial sustainability emanate, try to identify which interests underlie reform programs promoting them, and identify what this means for water operators in their day-to-day operations. The debates around possible (in)compatibilities or trade-offs between the two goals were useful in allowing students at IHE Delft to place their daily routines in a broader perspective. As the majority of our students are employed in water utilities or governmental agencies supporting the development of water services in their home countries, they appreciate the difficulties of reconciling these objectives. Through the exchanges during the course they realize they are not alone in facing such issues. However, as much as the students are able to engage with these debates, these discussions often do not provide them with actionable answers to questions on how to practically deal with such mixed mandates. While the discussions enrich their views on their work, these discussions do not necessarily make their job in reconciling social and commercial objectives any easier. I have become used to students asking us to share more experiences of how things 'are really done', what 'solutions' exist, what better operationalizations are out there, or what interventions can they learn from? Many of them are interested in examples and empirical case material on how water operators address and struggle with the challenges of reconciling, in so far as possible, social and commercial objectives. Many of them are themselves struggling to reconcile governmental policies that put pressure on their organizations to realize the Sustainable Development Goal Target of achieving universal service coverage by 2030, whilst simultaneously demanding that they do this in a financially sustainable manner.

To most of our students it is of little value to question the objectives and underlying argumentation of commercialization except for their personal intellectual interest. This may partly be because these ideas have become so deeply rooted in their work that these principles have become the boundary conditions within which they operate. They have become unquestionable and self-evident. This is the reality within which they have to function and make decisions. Questioning or radically changing these boundary conditions is not within their realm of influence. What they can perhaps do is try to adapt, change, tinker with, or accommodate, within certain boundaries, the implementation of models of service provisioning adhering to these principles in order to make it better suit their particular contexts.

The documentation and analysis of what models *do* once they are implemented provides the starting point of this thesis. I am interested in better understanding how these models become operationalized in the actual routines and practices of water utilities: how, if at all, do they guide or shape the behaviour, decisions and actions of utility operators and managers? Ultimately, I hope that this thesis will provide the students at IHE Delft, as well as a broader audience of water operators, with empirical analyses that can, at some point, help them do their job better.

3. Research objective and research questions

This thesis examines how water operators implement commercial principles in their daily operations. This research departs from the premise that if the reconciliation of social and financial objectives is challenging it must necessarily mean that during implementation there will be some degree of tinkering with one or both of the two principles of commercialization. This research documents, what I call, the operationalization of the model. Operationalization entails navigating, stretching and/or re-interpreting the boundaries of the policy model in such a way that it becomes operationally possible to claim the achievement of set performance objectives. This definition of *operationalization* takes inspiration from the rich literature on policy implementation and policy translation studies, which focuses on the study of implementation – with necessary interpretations and adjustments in the roll out of policies - as integral part of the policy process (see also Peck and Theodore, 2010). This research analyses the alterations visible in the daily implementation of the models and, in doing so, gains a better understanding of how utility operators deal with the trade-offs and tensions between expanding services and financial sustainability.

To achieve these objectives I have formulated the following research questions which guide the research presented in this thesis:

1. Which principles and arrangements for water provisioning prevail in the policy prescriptions emanating from the global water community for the provision of water services, and why?

This question elaborates on the policy prescriptions concerning different organizational models for water service provision, discusses the principles of commercialization in the water services sector and examines how these have become embedded within the organizational models.

2. How do operators translate the principles of commercialization into operational realities?

This question investigates how water providers operationalize the principles of commercialization in their daily operations and seeks to identify to what extent they comply with, or deviate from, these principles.

In answering these questions, I aim at establishing a dialogue which allows the model of commercialization, as currently being prescribed, to be better informed by the practical realities of its operationalization. The analysis of how the models are operationalized in the everyday should thus allow for a sincere reflection of the model of commercialization as it is currently being embraced and promoted by the ´global water community´.

Chapter 2: Commercialization in Water Services Policy Models[3]

As highlighted in the introduction, the international policy debate circling around possible and more desirable ways to organize the provision of water services has largely been held in ideological terms. As a result, discussions in the water sector geared at improving service coverage and sustainability have been focusing on the principles of any desired model or approach, often centered around the pros and cons of privatization, rather than zooming in on the details of actual water provisioning experiences. I use this chapter to show how commercialization of water utilities, in spite of many critiques, has become the policy paradigm in the water sector. I do this by tracing the historical developments leading to the gradual consolidation of the principles of commercialization, as they turned into the accepted norm for water service provisioning. The chapter ends with a brief presentation of how the principles of commercialization are incorporated in well-recognized approaches to water services.

1. Water provision under direct government control and Modern Infrastructural Ideal

In most developing countries, initial responsibility for water service delivery following independence from colonial powers was assumed by public sector agencies. In most newly-independent states, governments embraced the idea of directly providing basic necessities for its population as part of their country-building agenda (Pitcher, 2002). The idea that water services provision should be a public responsibility resonates with the 'ideal' bureaucracy idea of Max Weber, which postulates that services with a great impact on society should be provided by the public sector (Bozeman, 2004). Indeed, "that water was to be 'public' went largely unquestioned" (Bakker, 2013b:285). In such an arrangement a municipality or any other public governmental agency directly provides services (Braadbaart et al., 1999).

---

[3] This chapter is partially based on: Schwartz, K. and Tutusaus, M. (forthcoming). Legal Frameworks and Water Services: A case of Confused Identities. In Dellapenna, J. and Gupta, J. (eds). Elgar Encyclopedia on Environmental Law. Volume: Water Law. Edward Elgar Publishing.

While the water utility or water department may enjoy some degree of autonomy in this approach, it remains firmly embedded within the public sector and is part of the "government apparatus" (Braadbaart et al., 1999:7). At that time, the ideal organization for the provision of water services would adhere to the 'modern infrastructural ideal' (Graham and Marvin, 2001). This ideal states that service provision should be organized through "monopolies, for singular and standardized technological grids across territories" (Graham and Marvin, 2001:91). The modern infrastructural ideal thus envisages a single water utility providing water through a standardized, centralized water infrastructure network, which provides services through in-house connections to all inhabitants. To make such a system financially feasible and to allow all users, irrespective of their income, to connect to the network, government subsidies and cross-subsidies (within the operations of the utility) were considered acceptable and normal. Cross-subsidies entail charging higher fees to those consumer categories that are better able to pay for water services, whilst reducing the price charged to low-income consumers. Often this system of cross-subsidisation also incorporates an increasing block-tariff which links the price charged for water to the level of consumption by the consumer: the more water is consumed, the higher the price charged by the water utility[4]. In this way, the modern infrastructural ideal often incorporated an explicit redistributive component, with solidarities funded either through the national budget or through a differentiated tariff system (cf. Jaglin, 2008:1905). The many challenges experienced while pursuing the modern infrastructural ideal were mainly attributed to the limitations of the water infrastructure available for providing water services. Hence, achieving universal coverage was considered to be mainly a question of developing this infrastructure and extending the grid. "[I]f a utility would be technically equipped to provide services, the utility would do so" (Schwartz, 2008:49). Of course, the availability of financial resources is an intrinsic part of this technical challenge, as the required development of infrastructure to achieve the ideal would require considerable investments.

During this era of direct public management and the infrastructural ideal, the challenge of expanding the provision of water services was largely viewed as a technical issue. With the adoption of more sophisticated infrastructure and technology, low-income countries would be able to achieve the same infrastructural model that characterized water services provision in high-income countries. For example, in the city of Lagos, urban planners at the end of the 1970s

---

[4] The topics of cross-subsidisation and increasing-block tariffs have been subject to considerable debate and discussion. The use of such subsidies and block tariffs has been criticized as they often fail to target low-income consumers (but rather subsidise middle and high-income consumers) (Komives et al., 2005; Fuente et al., 2016)

"anticipated that within 20 years the entire city would be connected to a modern water supply system" (Gandy, 2004:368). In line with this thinking, the professional and academic literature on water services provision in developing countries in this era was mainly geared towards, on the one hand, presenting and discussing technological advances and, on the other hand, discussing financial arrangements to mobilize the needed investments.

2. The Drinking Water Decade (1981-1990) and alternative technologies

As touched upon in the previous chapter, the lead up to the Drinking Water Decade at the end of the 1970s[5] brought together the international water community around the ambition of ensuring universal access to water services. The Decade did indeed lead to increases in access to water supply, with estimates ranging from 1.2 billion (Economist, 1994) to 1.5 billion people (WHO, 1991) receiving access to water. However, much of the gains were offset by population growth. By the end of the Decade as many people lacked access to water supply services as at the start of the Decade (Economist, 1994). In line with the modern infrastructural ideal, efforts during these ten years focused on investments in water infrastructure. However much of the infrastructure built during the beginning of the 1980s was no longer functioning by the end of the decade, as maintenance of the water system had been neglected (WHO, 1991). The disappointing results of the Drinking Water Decade in conjunction with a wider tide of neo-liberal reforms in the 1980s and 1990s helped fuel a questioning of the "direct public management" approach and of the modern infrastructural ideal.

As part of this questioning, a discussion emerged around the *appropriateness* and *affordability* of the technologies[6] used for extending service provision. As a WHO (1991:7) evaluation report argued: "[…] the Decade witnessed a move away from oversophisticated[7] systems towards the use of technologies operated and maintained on a sustainable basis". The term 'appropriate technologies' was used to refer to low-cost technologies such as standpipes and water kiosks. According to its promotors, the word appropriate in the term 'appropriate

---

[5] These meetings were the Habitat: UN Conference on Human Settlements in Vancouver in 1976 and the UN Water Conference in Mar del Plata in 1977.

[6] The initial introduction of appropriate technologies in the water services activities of the World Bank dates back to 1976, when Kalbermatten was allowed to initiate a research project into "viable investment alternatives for low-income areas" (Black 1998: 7).

[7] Oversophisticated systems refers here to centralized networks and in-house connections

technologies' particularly refers to how well technologies match the demands and needs of low-income consumers[8] (WHO 1991). Although the WHO celebrated the move away from oversophisticated systems, it was clear that the less sophisticated technologies for supplying water were only targeting a very specific consumer category: the poor, or those living in low-income areas. Hence, with the invention of appropriate technologies, also a specific customer group was identified and brought into being. Centralized networks providing in-house connections remained the 'super-technologies of the rich', whilst appropriate technologies were deemed good enough for low-income areas (Gerlach and Franceys, 2010). As such, the increasing popularity of appropriate technologies in the 1980s and 1990s was not a shift away from the modern infrastructural ideal, but rather a partial replacement of the ideal for a particular category of society.

The shift prompted by the appropriate technology movement led to another development in thinking about water supply and sanitation. With the possibilities to provide services through appropriate technologies, the hardware part of providing services was no longer seen as the main bottleneck. This created room for focusing on issues around the 'software' of service provision (Cairncross, 1992; Ghosh and Cairncross, 2014). Hence, the New Delhi Statement[9] of 1990, which was to provide guidelines for the water services sector after the Drinking Water Decade, not only highlighted a focus on the use of low-cost appropriate technologies, but also included principles for institutional reforms, community management and sound financial practices.

3. Shifts in the role of governments

At the time of the New Delhi Declaration, public water utilities in many developing countries showed a dismal performance. Service coverage frequently did not extend beyond 50% and non-revenue water levels of 60% or higher were no exception (Panayotou 1997; Nickson 2002; Mwanza 2004; Mwanza 2005). Often the utilities were overstaffed, having about six times more employees than the 5 staff per 1,000 connections deemed appropriate by development agencies (Haarmeyer and Mody, 1997). Moreover, prices of water did not reflect actual costs,

---

8 Research about the satisfaction of users regarding access through these technologies is not conclusive and points in different directions. A study of service provisioning in low-income urban areas in Kenya found that consumers had the highest level of dissatisfaction with water services provided through water kiosks. "Although people are forced to rely on them, they do not like them as options for water supply" (Gulyani et al. 2005:1262).

9 The New Delhi Statement came forth from a Global Consultation on Water Supply and Sanitation

water utilities lacked capacity and were subject to political interference (Baietti et al., 2006; Bakker, 2003; Wami and Fisher, 2015). As analyses of that time pointed out, these utilities could only continue operating thanks to large government subsidies (World Bank 1994; Idelovitch and Ringskog 1995; Foster 1996). Problem diagnoses attributed these performance weaknesses to the fact that water utilities were typical bureaucratic organizations. These were believed to be intrinsically unable to operate efficiently and effectively (Panayotou, 1997) because of perverse incentives and inadequate accountability structures. Utilities were rewarded for meeting administrative targets and accountable to governments, rather than being rewarded for the actual quality of the services they delivered and being accountable to their customers.

A similar dissatisfaction with the functioning of bureaucratic organizations sparked a broader wave of public sector reforms in the world. Starting in New Zealand, the United Kingdom and the United States in the 1970s and early 1980s, a large number of countries initiated reforms to shift the role of the government from "a producer to an enabler" (Hughes, 2012). Although the exact nature of the resulting public sector reforms differed from one country to the next, they all shared the introduction of more market-orientation and output-based forms of accountability, as well as an increased emphasis on user-orientation by adopting 'business' style measures (Schwartz, 2006). With governments embracing their new role of enabler, the role of 'non-state actors' became more important: the provision of public services was (partly) delegated to non-state actors, be it private companies, parastatal organizations or community-based organizations (Moriarty et al., 2002; Scott, 2008). "At all layers of government, and almost on a global scale, public officials embraced or succumbed to an 'away with us' attitude" (Ringeling, 1993 in Hill and Hupe, 2002).

4. The Privatization Decade: 1993-2003

Although the private sector historically played an important role in the water services sector of many countries[10], until the 1980s its role as a direct water provider was limited. The public sector reforms of the 1990s, however, entailed that donors and development banks started advocating for more involvement of the private sector in providing services. With the

---

[10] In many countries the initial development of the water services sector in the 19th century was a private sector initiative catering to high-income households that were willing to pay for better quality water (Blokland, 1999; Furlong, 2009)

establishment of a concession contract in Buenos Aires in 1993, the first of a series of large public-private partnership contracts, the "privatization decade", which lasted until 2003 (Franceys, 2008), got off to a start. During this decade, a large number of countries adopted new water policies and laws to allow for the involvement of the private sector in providing water services. These include Senegal, (1995), Uganda (1998-2001), Mozambique (1999), Indonesia (2004), Mexico (1992), Namibia (2000), Zambia (1994-2001), Kenya (2002), Tanzania (2002) and South Africa (1998)[11].

The privatization decade sparked off an intense debate on how to best organize water services provision, focusing on which incentive structure would make operators perform well. The reforms touched upon both the organizational and the financing elements of service delivery. The International Development Banks and most donors favored private sector organizations as service providers, believing that these would operate more efficiently, while also remaining outside the influence of politicians and government agencies. The idea was that this managerial autonomy would accompany financial autonomy, with private water utilities being expected to operate on the basis of full-cost recovery. This, in turn, would make it possible to reduce or eliminate government subsidies for water services provisioning. Although cross-subsidies and increasing block tariffs would still be possible (and were abundant) within this set-up, the credo was that water utilities had to become financially autonomous.

Private water providers thus were increasingly regarded by governments and donors as long terms partners in extending and improving water supply services (Mehta, et al., 2007). An underlying discussion, prominently articulated by the World Bank, concerned the appropriate incentive structures under which water utilities should operate, and about the 'institutions' which shape these incentives. The 1994 World Development Report, for example, concludes that "numerous examples of past failures in public provision, combined with growing evidence of more efficient and user-responsive private provision, argue for a significant increase in private involvement in financing, operation, and – in many cases – ownership" (World Bank, 1994:8). Apart from more ideology-inspired arguments for involvement of the private sector, developments in England and Wales in 1989 also had a serious impact on the momentum of private sector involvement. In 1989, the English Government privatized the English and Welsh water supply sector through divestiture. In doing so, they demonstrated that privatization in the

---

[11] In addition to catering for private sector involvement, the water policies developed in this era also strongly reflect the principles of integrated water resources management.

water services sector was possible and that it could be done at a previously unprecedented scale.

5. *Promoting privatization*

The literature on the management and governance of water services management in this era can largely be divided in two main strands. A first strand of literature firmly promoted private sector involvement. This policy-oriented literature originated in international development banks. It combined rather generic assumptions on the benefits of private sector involvement with a very small selection of empirical cases that were to demonstrate that the private sector could deliver. The argumentation rested on the assumption that the managers and employees of a water utility behave as rational beings: they were supposed to act according to known incentives that are shaped by the organizational models. This assumption about the behavior of utility managers and employees was not informed by empirical studies of the actual practices and doings of these individuals, but deduced from the theoretical (and ideological) principles of the model. Hence, the theory predicted that a private operator would be more motivated than a public one to pursue efficiency gains by maximizing revenue generation and reducing excessive costs.

The cases repeatedly referred to in support of this theoretical prediction included SODECI in Ivory Coast (Lewis and Miller 1987; Haarmeyer and Mody 1997; World Bank, 1997;), La Paz-El Alto (Komives, 1999; Lee and Floris 2003), Buenos Aires (Idelovitch and Ringskog, 1995; Lee and Floris 2003) and the affermage contract in Senegal (Brocklehurst and Janssen 2004). Each of these cases, which appeared and re-appeared in various publications and presentations at international fora, contained specific lessons for successful private sector involvement. SODECI represented an example of private sector involvement that dates back to 1957. It was used to illustrate that privatization was a tried and tested long-term solution. Moreover, as it concerned a case in Sub-Saharan Africa, it underscored that privatization is also an option for less-developed countries. The privatization of Buenos Aires represented the first large-scale privatization in a middle-income country in the 1990s. It, thus, showed how privatization did not need to remain limited to countries like England and Wales but could be applied in developing countries as well. The relevance of the La Paz-El Alto case was that the tendering of the concession contract took place on the basis of service coverage (rather than tariff). As the private operator who was awarded the contract committed itself to universal service coverage, the case countered criticism of cherry-picking (Swyngedouw, 1995) and illustrated that privatization could also benefit the poor. The affermage contract in Senegal again represented an example of privatization in Sub-Saharan Africa. Moreover, the type of contract (affermage) and the organizational set-up of the sector (with an independent regulator monitoring the contract[12]) presented another innovative approach to privatization in Sub-Saharan Africa.

6. Critiquing privatization

Privatization in the drinking water sector did not go uncontested. A second stream of literature that emerged during the privatization decade strongly criticized the promotion of privatization, largely focusing on its detrimental impact on the accessibility of water services for low-income consumers. The critics of the privatization paradigm argued that there is little empirical evidence to support the belief that privatization leads to the expansion of services. They predicted that it would instead further increase inequities in supply and coverage (Castro, 2007; McDonald, 2002; Loftus, 2005; Bakker, 2010). This literature associated the involvement of the private sector with cherry-picking, in which utilities prioritize service delivery to those areas and customers where profits would be highest (Swyngedouw, 2005), low-expansion rates and the cutting off of services in low-income areas (McDonald, 2002). According to these critics, privatization would lead to a prioritization of efficiency - in the form of cost containment and revenue enhancement - over equity (Jaglin, 2002; Roa-García, 2014; McDonald, 2016; Marson and Savin, 2015).

Similar to the promoters of privatization, many of these critiques and critics make use of ideological-theoretical simplifications to explain or predict the behavior of water operators and their staff. The behavior of managers and employees in private operators is deduced from the model under which the water utility operates rather than basing it on empirical studies of the actual practices and operations of water utility staff. And similar to the pro-privatization literature, also critical scholars make repeated use of a limited number of case studies to make their argument. An example of such a case is the Cochabamba concession contract in Bolivia, which following a period referred to as the 'Water Wars', was cancelled in 2000 (Assies, 2003; Schwartz and Schouten, 2007). The critical discussions of this case often emphasize the

---

[12] The use of both an independent regulator and delegated management through a contract essentially combines two regulatory traditions (the French model of regulation and the English model of regulation). For a more elaborate discussion of these traditions see Foster (2005).

significant tariff increases of up to 200% (Spronk, 2007) to illustrate how privatization leads to water services that are only affordable to middle and high-income households.

Representatives of trade unions and civil society organizations manifested themselves as important members of those critical of water privatization in this era. Many of them, in alliance with or represented by organizations like Public Services International, the Transnational Institute and the Corporate Europe Observatory, helped produce a substantial body of grey activist literature produced to counter privatization. A review of this literature reveals two main lines of argumentation. The first is the principled critiquing of the privatization model, often based on the use of selected case studies that illustrated the negative consequences of privatization (Hall, 1999; Bayliss, 2003). The second line of argument is the presentation of successful alternatives to privatization. Under the heading of ´remunicipalization´ (Pigeon et al., 2012; Lobina, 2005) or ´public alternatives´ (Hall et al, 2013) the activist-scholars provided selected case examples to demonstrate that it is desirable to keep water services provision in public hands.

Although the literature streams described above have contradicting arguments, both build their arguments in favor or against the suitability of a particular model of services provision mainly on ideological and theoretical grounds. Hence, their arguments focus on promoting or critiquing an ideal or the principles underlying a model, rather than zooming in on the actual working of the model in practice. Where empirical cases are used to support an argument, these tend to assume that the providers behave in accordance with what the model predicts, and that incentive structures are indeed what the model or theory says they are or should be. In other words, in much of the debate the same theories that are promoted or critiqued were used to collect and explain empirical observations. How the model is operationalized in everyday practice becomes relevant in these writings only in so far as it supports the claims that the authors want to make. In doing so, this literature produces very little insight into the operationalization of the models in practice, or does little to explain the actual behaviors and incentives of utilities and their staff. In this way, it also provides little guidance to utility staff and operators as to how they best can operationalize the model in their specific context.

7. The end of the promise of privatization

The early 2000s saw a notable shift in the momentum for private sector involvement in the water services sector. On the one hand, there was a more general backlash against the privatization of public sector services. Although water policies developed in the 1990s allowed for and stimulated the introduction of privatization, there were few utilities that were actually privatized. Moreover, studies analyzing the impact of private sector involvement did not yield much convincing evidence to support the belief that privatization reforms would yield cost savings and efficiency gains (Boyne, 2002; Bel et al., 2010; Domberger and Jensen, 1997; Jensen and Stonecash, 2005; Marsh, 1991). In addition, and perhaps as an outcome of the earlier debates, privatization - especially the outsourcing of public services - was increasingly viewed as being narrowly efficiency-oriented rather than service-quality oriented. As a result, citizen and government engagement were now again promoted as important dimensions of water services provision (Box, 1999; Christensen and Laegreid, 2002, Denhardt and Denhardt, 2000; Nalbandian, 2005).

At the same time, the appeal of large-scale privatization contracts began to fade because the few highly-publicized cases increasingly revealed operational difficulties. During the privatization decade, a main focus for introducing the private sector into water service provisioning was "their supposed ability to supply private money" (Marin 2009:138). However, an extensive review by the World Bank indicated that the expectation that the private sector would finance water services provision had been too optimistic. According to the review, the main contribution that can realistically be expected from the private sector is the improvement of operational efficiency and service quality (Marin, 2009). Also others showed that the anticipated influx of private sector capital in the water services sector remained far below expectations (Schwartz, 2008). In addition, the East-Asian financial crisis of 1997 caused enormous challenges for the concession contracts in Manila and Jakarta. In Sub-Saharan Africa, ambitious privatization projects failed to materialize (Ahlers et al., 2013). Concession contracts that were presented during the 1990s as guiding examples for the development of the water services sector, were terminated under civic pressure. Following the collapse of the Cochabamba concession contract in 2000, the Buenos Aires concession contract that marked the start of the ´privatization decade´ over a decade earlier was also terminated in 2005. Soon after, the collapse of the concession contract in La Paz-El Alto (2005) signaled the definitive end of the privatization decade.

Yet, according to a senior staff member of the World Bank: "[t]he story on private sector participation is not over, it's simply being rewritten. When the emotions over this issue subside over the next few years, there will be a robust return to private sector participation that involves subtler forms of risk sharing" (Ahlers et al., 2013). What this quote highlights is both that private sector involvement remains an important management option for the World Bank (see also Marin, 2009) and that this private sector involvement primarily is about operating water services rather than investing in water infrastructure. Rather than a source of funding, the international funding community now views the private sector as an operator of services, focusing primarily on involving private sector actors in operating and maintaining water systems. The consensus now seems to be that whereas operational and commercial risks can be attributed to the private sector, investment risks should largely remain with the public sector.

8. *The principles of privatization remain: Commercialization*

Despite the discontent with the involvement of the private sector, the idea that a water utility should act as a financially and managerially autonomous entity and pursue financial efficiency remains firmly entrenched in the thinking of many international water sector experts. Rather than privatization, the term used is commercialization. While privatization referred to a shift in ownership from the public to the private sector, commercialization refers to a "reworking of the management institutions (rules, norms and customs), and entails the introduction of markets as allocation mechanisms, market simulating decision making techniques and the displacement of Keynesian-welfarist by neo-liberal principles in policymaking" (Bakker, 2003: 331). Hence, although the end of the privatization decade signaled a renewed engagement with public water utilities by the international water community, many of the private institutions promoted during the privatization decade survived and became part of this engagement. Illustrative of this new consensus is a publication from 1999, which, under the title of *Private Business, Public Owners* (Blokland et al., 1999), presents a water services model of publicly owned but commercially operating water utilities. In this model, public water utilities were expected to operate on the basis of commercial principles. This shift was grounded in the belief that "[w]ell-run public utilities of the developing world have much in common with efficient private providers" (Marin 2009:147). What utilities needed was the introduction of private sector 'discipline' to improve their performance (Furlong, 2015). This form of 'commercialization' was considered more favorable and less controversial than models involving private sector organizations (McDonald, 2014).

Table 1 *Different definitions of Commercialization according to different authors*

	Boyne, 2005	Bakker, 2007	Hughes, 2012	Kitonsa and Schwartz, 2012	Furlong, 2015	Moriarty et al, 2013
Commercialization	"Property right" dimension combined with owners right of profit is argued give for incentives the owner to monitor and control the organization for better performance and maximize the efficiency of the organization.	Proponents of commercialization argued that water as scarce resources must be priced at full priced (economic and environmental cost) thus has to be managed by profitably "private" companies believed has more direct accountability to its client (customer and shareholder) which imply relatively free from political domain.	Commercialization is argued can resulted in reducing the public expenditure, improve efficiency in public finance and will lead government to adopt a more flexible management system as characterized the private sector.	• Full cost recovery operation of water utilities; • Water utilities as autonomous entities; and • Performance target to enhance accountability of water utilities, assessed by independent parties.	Commercialization marked by "deregulation and reregulation" in order to promote private sector participation (PSP) to achieve full cost recovery and independence from local government.	Financing of all the cost of service and often with combination source of financing.
Principles	• Performance control • Efficiency • Profit orientation	• Full priced (cost recovery) • Profit based • Direct accountability • Autonomous entities	• Cost Efficiency • Flexible management	• Full cost recovery • Autonomous entities • Performance management	• Full cost recovery • Autonomous entities	• Full cost financing • Combination Source of Financing

Different authors have provided different definitions of commercialization. Table 1 provides an overview of these different definitions. Despite some minor differences, there seems to be considerable agreement on the main principles of commercialization (Boyne, 2002; Bakker, 2007; Hughes, 2012; Kitonsa and Schwartz, 2012; Furlong, 2015; McDonald, 2014). All definitions emphasize financial enhancement and incentives to operate more efficiently. In this thesis I consider commercialization to be characterized by two principles: a commercialized water utility should operate on the basis of full-cost recovery and is able to operate as a fully autonomous entity.

The principles of commercialization represent a set of guiding actions according to which water providers should organize their functions and operations. These principles are promoted and manifest themselves in a variety of possible organizational and service provision models ranging from full privatization or corporatization, to the adoption of commercial principles in the management and operations of a public organization. Commercialization does not require a change or replacement of the type of organization, but refers to "a change in the way the organization operates and is managed" (Kitonsa and Schwartz, 2012: 181).

a. Commercialization in Community Management and Small-Scale Private Operators

Initially, the proponents of private sector involvement and the proponents of appropriate technologies formed relatively distinct groups within the domain of the water services sector. Discussions on private sector involvement mainly revolved around large urban centers, which had a sufficiently large and wealthy market to interest private water operators (Moriarty et al., 2002). As such, it is not surprising that the main concession contracts established during the initial years of the privatization decade concerned large capital cities (Buenos Aires, Manila, Jakarta, Johannesburg, La Paz-El Alto) or large primary urban centers. Groups advocating for appropriate technology were more focused on peri-urban areas and towns which were not serviced by a water utility. Appropriate technology thus became the approach for (parts of) urban settlements where the modern infrastructural ideal was difficult to realize, a method to ensure that these unserved areas would also receive access to water services. As such, the two approaches – privatization and appropriate technology - often targeted different water service realities, different parts of the city, and different types of consumers. However, particularly during the second ('soft-ware') phase of the appropriate technology approach, certain similarities between the two groups became apparent. The shift from a technology focus to one incorporating institutional dimensions of service provision shifted the focus of both streams towards organizational arrangements which ensured autonomy from the government apparatus and a focus on recovering (at least) operation and maintenance costs.

Community involvement in the provision of water services is supported by two, rather distinct, frameworks. The first highlights reduced involvement of the government apparatus and the incentive structure that community members have to ensure efficient water provisioning. Community involvement is then seen as a positive factor as long as the members of the community understand that it is in their own interest to operate and manage a water system (Carter and Howsam, 1999). Community management thus promotes autonomy, goals of efficiency and full cost recovery, which are aligned with the principles of commercialization. In examining community participation, Cornwall and Gaventa (2001:10) conclude that "[t]he neo-liberal paradigm emphasizes user participation, not only in project design but also in bearing the cost for service delivery". The second framework presents an argumentation of empowerment as the rationale for community involvement. In this framework the voice of community members in deciding about the provision of water services is emphasized (Harvey and Reed, 2007). It revolves much more around democratization and the participation of community members and is less concerned with economic objectives of efficiency. As highlighted by Knippenberg (1997:15) "[...] consumers become active partners whose voices count. Beyond improving efficiency through increased transparency and reduced leakage, the community's full partnership in the decision-making process and the fact that the poor become better organized and more vocal in demanding [...] services, improves effectiveness and financial viability". Although community management remains highly promoted in the water services sector for the operation and maintenance of decentralized systems, various authors have criticized the overly idealistic portrayal of communities in such arrangements (Cleaver, 1999; Manor, 2004; Rusca et al., 2015).

Likewise, the re-appraisal of small-scale providers fits within the search of alternative management arrangements for locations outside large urban areas. Small-scale providers are particularly promoted in rural areas and small towns where the existing conditions make community management difficult, or in peri-urban areas where larger utilities cannot reach (Kleemeier, 2010, Njuri, 2006). Small private operators are argued to offer a suitable alternative for water supply based on their claimed accountability, and their capacity to design and deliver 'public' services (Schacter, 2000). Despite the lack of thorough documentation on how these private operators function and react to the realities of small towns (Kleemeier, 2010), much donor-led literature, such as that deriving from the World Bank, Water and Sanitation Program (WSP) and Public-Private Infrastructure Advisory Facility (PPIAF), speaks favorably of the introduction of the local private sector in these settings (World Bank, 2017, WSP, 2011; PPIAF, 2009). This follows from a wider acceptance of small-scale operators as legitimate actors to expand services as they would be able to operate without the support of the government and have the intrinsic motivation to operate efficiently (Schaub-Jones, 2009). In increasingly complex setting such as small towns, private sector operators are argued to be best suited to cope with the increasing demands for higher level services deriving from economic and population growth in these settlements (Lockwood and Le Gouais, 2011).

9. Three dominant models of service provisioning

In contemporary policy thinking, there seems to be widespread consensus on the possibilities available to provide and expand drinking water services. A comparison of the literature produced by multilateral donor organizations, national development agencies, advocacy organizations and academics reveals a striking similarity in the range of options these propose and discuss. These options are delegated management (Prasad 2006; Marin 2009), commercial public utilities (Blokland et al., 1999; Smith, 2003; Schwartz, 2008) and community management (Moriarty et al., 2013. Adank, 2013; Pilgrim et al., 2007). The thinking and understanding of how the sector can and/or should be organized, therefore, happens largely through discussions about (variations of) these three options.

Table 2 Three dominant models

Model	Organization	Finance	Infrastructure
Delegated Management	Assets: Government agency Operations: International or national private operators (through contracts)	Investment: Government and donors Operations and Maintenance: Cost recovery	Inhouse connection or yard taps for those in the areas where existing network exists, and other technologies such as kiosks, standpoints (prepaid meters) in low income areas

Model	Organization	Finance	Infrastructure
Commercialized Public Utility	Assets: Public water utility Operations: Public water utility (through government statute)	Investment: Public water utility, Government and donors Operations and Maintenance: Cost recovery	Inhouse connection or yard taps for those in the areas where existing network exists, and other technologies such as kiosks, standpoints (prepaid meters) in low income areas
Community Management	Assets: Community-based organization Operations: Community-based organization	Investment: Government and donors Operations and Maintenance: Cost recovery	Traditionally point sources (kiosks, standpipes, etc.), but increasingly yard taps.

10.   *Adjustments to the model as entry points*

In much of the literature on water services, adjustments to the models are sometimes proposed or tolerated, but only as temporary measures to help achieve the original ideal of a commercial utility that operates at arms-length of the government (see for example Baeitti et al, 2006). Rather than as temporary deviations of something that is more desirable or better (the model), in this thesis I consider possible adjustments during implementation deviating from the model ideal as an important entry-point for describing and explaining how utilities work and function. Hence, I see the modifications that operators make when trying to implement a model as pointers to understanding why they do what they do. I thus consider these adjustments – the modifications, experimentation, tinkering – as intrinsic to the functioning of water services provisioning, as something that is needed.

In order to theoretically anchor this hypothesis, I resort to different strands of policy implementation theory (Chapter 3). I use a review of the policy implementation literature to learn about how the adjustments to policy ideas or models that occur during the implementation can be understood and how they can be explained.

Chapter 3: Policy Models – Formulation, Implementation and Change

In the previous chapters I have presented the three dominant models currently being promoted and implemented in the water services sector and I place these in the historical development of thinking about and doing water services provision in the international policy domain. In this chapter I present and discuss ways of understanding policy formulation, implementation and change, in search of theoretical inspiration for addressing the concerns and questions outlined in the previous chapters. These concerns center on the usefulness of the three dominant models for the operational practices of water utilities, and more broadly of the language of commercialization in which they are accommodated. Although very similar models have been promoted for years already, results in terms of improvements in water services provision and extending access are disappointing. Yet, these disappointing results have not done much to shed doubts on the models themselves.

Policy implementation theories provide a useful source of inspiration to look at the application of commercial models of water service provision as policy implementation theories largely encompass the interface of formulated policies with the practical everyday manifestation of those policies. There are different approaches to the study of policy implementation. A prominent one is to look at policy implementation as a consecutive step following policy formulation. In this approach, policy implementation consists of the neat execution of a policy idea or model. Successful implementation then means that the implemented reality closely mirrors and matches the original policy idea. Hence, implementation studies using this conceptualization consist of comparing the original idea with the implemented reality. If differences between the two are identified, it signals implementation failures that prompt proposals for improving the implementation process or redressing the conditions on the ground so that these allow for the better and smoother realization of the policy. Most policy and academic writings on water service provisioning adhere to this conceptualization of policy implementation. One important effect of this conceptualization is that the policy ideas or models themselves remain outside of that what is questioned, which also means that these writings do not contribute much to policy learning (change).

In this thesis, I am interested in using the difference between implemented realities and policy models to (also) interrogate the models themselves. By understanding policy implementation from the perspective of those doing the implementation, taking their agency and creativity seriously, I hope to contribute to a different understanding of water services provisioning. More broadly, I am interested in identifying possibilities for mutual learning between those designing policy models and those implementing them. To do this, it is useful to adopt a conceptualization of policy implementation that accepts differences between implemented realities and policy ideas or models as legitimate (and perhaps necessary), seeing them as the adjustments and adaptations needed to make a model function in a complex reality. In this chapter, I further review and discuss scholars who have elaborated on this conceptualization of policy implementation and the policy process.

1. Policy development: a linear process

The most influential conceptualization of the policy process is what Sabatier has referred to as the "stages heuristic" (Sabatier 2007:6). It depicts the policy process as constituting of a series of sequential steps, each linked to specific institutional actors (Nakamura, 1987, deLeon, 1999). Although the exact identification of steps or their naming may differ between authors, there is a considerable degree of overlap between them in terms of their overall view of what a policy process consists of. Hence, Anderson (1975) presents the policy process as constituting of problem formation, formulation, adoption, implementation and evaluation, whereas Polsby (1969) speaks of policy initiation, incubation, modification, adoption, implementation and appraisal. Nakamura (1987: 143) who summarizes the stages approach in the three broad stages of policy formulation, policy implementation and policy evaluation, refers to these conceptualizations as the 'textbook policy process'.

In the textbook version, the policy process consists of a linear continuum of stages from policy development to policy implementation. In it, policy development appears as a rational and technocratic process involving the identification of a problem to be solved, followed by the proposal of solutions in the form of policy prescriptions (deLeon, 1988).

Figure 1   Variations of the Policy Process (Source: Nakaruma (1987); Polsby (1969); Anderson (1975))

Nakamura (1987)		
formulation	implementation	evaluation

Laswell (1963)		
intelligence → recommendation→ prescription→ invocation →	application →	appraisal → termination

Polsby (1969)		
policy initiation→ incubation → modification → adoption →	implementation →	appraisal

Anderson (1975)		
problem formation → formulation → adoption →	implementation →	evaluation

Jones (1977)		
problem identification → program development →	program implementation →	program evaluation → program termination

Although popular and influential in policy development and policy planning, this text-book version of the policy process is of limited use for describing and understanding actual policy processes. For one, it does not explain (and predict) how one stage leads to another (Nakamura, 1987). Furthermore, its rather legalistic focus makes it difficult to recognize the inter- and intra-governmental relations that exist beyond the rule of law and the prescribed roles and responsibilities for each of the involved parties. Lastly, it provides little space to integrate learning in the policy cycle (Sabatier and Jenkins-Smit, 1993). Yet, the conceptualization of the policy process as consisting of linear stages continues to be influential (deLeon, 1999), perhaps because of the normative attractiveness of the idea that policy is made by elected bodies and subsequently carried out by unelected officials (Cairney, 2011).

The 'textbook' understanding of the policy process assumes that it is possible to control the nature and direction of policy implementation (Maynard-Moody et al., 1990). It is a rather top-down (Matland, 1995) perspective to policy implementation, defining it as being concerned with the "degree to which actions of implementing officials and target groups coincide with the goals embodied in an authoritative decision" (Matland, 1995:146). It is a conceptualization that is often used to identify whether "legally-mandated objectives were achieved over time and why" (Sabatier 1986:22). The ultimate aim of analyses in this school of thought is to help improve implementation, which means making implementation outcomes coincide with policy goals. This view on policy implementation puts particular attention to the views and needs of those that are in charge of policy formulation. It seeks answers that would help validate the original policy premise and, as such, would not necessarily question the mismatch between implementation and formulated policy as an issue of the policy formulation itself (problem and solutions definition), but rather as an issue of implementation. In this conceptualization, successful policy implementation means that original policy goals are achievable and that there is consensus between policymakers and implementers (Hill and Hupe, 2002).

2. *More actors and less linearity*

The stages heuristic makes it difficult to recognize and incorporate the participation of non-state actors in the policy process. It limits the identification of relevant policy actors to the elected officials and policy makers who develop policy and the civil servants who implement it. This obscures the propensity that policy processes are often marked by many more actors, who come together in complex networks and who often have different interests (Hill and Hupe, 2002). In this case, when more actors are involved in the making of policies, policy outcomes would be negotiated outcomes. This questions the idea that policy making is an absolute and objective agreement among parties. Simon (1982) questions the idea of the policy process as consisting of stages informed by objective gathering of information, a design of alternative solutions, the choice of the preferred alternative and implementation that can be measured against clear objectives. Rather, policy makers act as satisfiers, seeking a satisfactory solution rather than an optimal solution. The fact that the process is not bounded to a rational reasoning, leads to policies that can be unclear or even conflicting[13] (Simon, 1982). Finally, the idea of a formal policy decision as the starting point for implementation ignores the historic development of the policy. "[T]he policy-formulation process gives implementers important cues about the intensity of demands, and about the size, stability and degree of consensus of those pushing for change. An analysis that takes "policy as given and does not consider its past history might miss vital connections" (Matland 1995:147). These are ideas that are particularly useful in this thesis as I question the paradigmatic promotion of the package of policy principles linked to commercialization as the way to sustainable water services

3. *Adjustments to policy during implementation*

In seeking explanations as to what operators do in their day-to-day activities to reconcile, adjust or comply with the policy ideal, I am particularly interested in an approach that acknowledges operators as policy actors. Changes to the policy idea or model that happen during

---

[13] Not only may the policy itself be unclear but a policy may also conflict with policies emanating from other, separate but related, governmental bodies (Cohen et al., 1972). An example of such governmental bodies concern the existence of Ministries of Water and Infrastructure alongside Ministries of Environment. Both such ministries are likely to produce policies relating to the water sector.

implementation are then not necessarily shortcomings, but reasoned adaptations and modifications that allow for better water provisioning to happen.

Theories of policy translation provide such a conceptualization. The term translation indicates the conceptualization's rejection of a simplistic causal linearity between policy formulation and implementation. Instead, it directs attention to how original policy ideas always require adjustments and changes to 'work' in practice (Mukhtarov, 2014). Hence, policy translation sees transformations and modifications as an integral part of policy development and implementation. Differences between design and implementation are not an indication of faulty implementation, but are expected and necessary. It is the process of mutation in which policies are molded for implementation in a particular context (Stone, 2012). Policy translation thus stresses that modifications are intrinsic to implementation (Mukhtarov, 2014).

b. Actors: implementers

This understanding of policy implementation in terms of translation underscores that implementation of policy models is part of the process of policy development. In this theorization, implementers are seen as actors that form an integral part of policy development, and are not simply relegated to the more passive and instrumental role of implementing models that have been formulated elsewhere. In translation theories, actors involved in implementation are important as they are the ones that actually make sense of the policy ideal. In this sense these 'functionaries' carry an enormous weight in co-shaping how projects are realized (Larner and Laurie, 2010). Policy translation is necessarily an actor-based approach as it gives a very important role to the practitioners who embody and manage the contradictions between policies and messy realities. It is through them that policies are transformed (Roy, 2012 in Larner and Laurie, 2010; Mukhtarov, 2014). Indeed, the actions and manoeuvres of the implementer take center-stage in the work of policy translation. Michael Lipsky's (1980) conceptualization of the 'street-level bureaucrat' marks the start of such a translation-like conception of the policy process, one in which the implementer is considered as an active agent in the policy process (Hill and Hupe, 2002). Lipski (1980) argues that "the decisions of street-level bureaucrats, the routines they establish, and the devices they invent to cope with uncertainties and work pressures, effectively become the public policies they carry out" (Lipsky, 1980: xii). By emphasizing the importance and power of implementers, Lipsky also questions the ability of policymakers to fully control and steer the implementers and the implementation process (Hill and Hupe, 2002).

Understanding the policy process in terms of translation resonates with what has come to be referred to as the 'bottom-up approach' (Hill and Hupe, 2002), as it looks at implementation from the perspective of those doing the implementation. The wider bottom-up literature emphasizes that public service workers have discretion to "not only deliver, but actively shape policy outcomes by interpreting rules and allocating scarce resources" (Meyers and Vorsanger, 2005:246). In these approaches implementers are seen as 'sociologically complex actors, located in (shifting) organizational and political fields, whose identities and professional trajectories are often bound up with the policy positions and fixes that they espouse' (Peck and Theodore, 2010: 170). Indeed, Lipsky "turns the policy-process upside-down by claiming that street-level bureaucrats are the real policy makers" (Winter 2012:214). This makes street-level bureaucrats very influential as they have considerable impact on the lives of citizens. "The way in which street-level bureaucrats deliver benefits and sanctions structure and delimit people's lives and opportunities" (Lipsky, 1980). In order to understand policy implementation, it is then required to understand the goals, strategies, activities and contacts of the actors involved in implementation (Matland, 1995). In a similar vein, for Mosse (2004), implementation is best understood as a dance between different actors representing different spoken and unspoken 'transcripts' and interests which give way to fragmented practices. This also permits bottom-uppers to suggest that adaptive administrations making use of the direct knowledge and expertise of the street-level bureaucrat about local conditions will be more successful in policy implementation (Maynard-Moody et al., 1990). However, these bottom-up conceptualizations may lead to an overestimation of the influence of implementers as completely independent actors in the policy process. In this thesis, I am interested in how implementers translate the dominant commercial models of water provisioning by adjusting them on the ground.

c. Contact zones and multiple policy domains

In order to study the adoption of policy models, it is important to identify how the making of the policy model is transferred to the place where it needs to become reality: who is involved in this transfer and by which channels. Maynard-Moody and Herbert (1989) suggest that there are two loosely coupled but distinct policy processes. One process is that of legislative policy making, in which an authorized government actor makes a policy decision. The second process concerns administrative policy making, which according to Maynard-Moody and Herbert (1989:139) involves "the entire range of public action, from agenda setting and innovation through implementation and evaluation". Relevant in this approach, is that Maynard-Moody extends the role of the street-level bureaucrats from adapting policy during the policy implementation stage to being responsible for an entirely distinct and independent policy process. In other words, rather than limiting the role of street-level bureaucrats to the implementation stage in the policy process, Maynard-Moody argues that these administrators essentially operate over the entire range of stages of a policy process. The fact that he also recognizes a legislative policy process then suggests the co-existence of multiple policy processes at the same time. Maynard-Moody argues (1990:140) that "it is important to examine both the distinguishing features of each arena and their points of connection". In that case, it is possible that implementation and policy design develop and evolve differently alongside each other. It is possible that these two domains develop potentially differently according to Maynard-Moody (1989) as they constitute and serve different purposes. "[A]dministrative policy making is dominated by ideas, norms, routines, and choices of non-elected public employees, whereas legislative policy making is dominated by the perspectives of elected officials" (Maynard-Moody, 1989:137). These levels of policy 'making' resemble the conceptualization of multiple domains of water politics proposed by Mollinga (2008). He distinguishes between domains of global water politics, hydropolitics between states (inter-state), the politics of water policy in the context of sovereign states, and the politics of the everyday (Mollinga, 2008). Similarly to Maynard-Moody (1989), these domains serve different purposes and "they have different space and time scales, are populated by different configurations of main actors, have different types of issues as their subject matter, involve different modes of contestation and take place within different sets of institutional arrangements" (Mollinga, 2008:12).

Despite the distinctive conceptualization of these levels or domains as if they could be seemingly decoupled, these authors also suggest that there are strong interactions between them. Maynard-Moody (1989:137) argues that "[t]he argument that the two processes are distinct does not deny their essential overlap". Mollinga (2008:13) identifies the linkages between the different domains as a separate fifth domain, which analyses "how policy issues and water contestations travel across the different domains". These overlaps or travels between domains or levels are a particular point of attention in theories of policy translation, where the points where levels or domains or processes meet are called 'contact zones' (Blaustein, 2015: 82). It is in these zones of contact between different actors that contradictions are negotiated. These contradictions manifest themselves in the difficulties to reconcile practices as defined in the policy model and the capacities of local practitioners to implement them as such. This space of negotiation brings together global and national networks and local practitioners (Cochrane and

Ward, 2012). What this means is that policy is not simply espoused by national agencies, but rather co-produced by local practitioners in these contact zones (Blaustein, 2015). The factors that influence the translation of policies include: path dependency arising from past decisions in a given location; institutional and structural impediments that exist in the location where the policy model is being implemented; and insufficient technological, economic, bureaucratic and political resources to implement a particular policy model (Benson and Jordan, 2011: 372). This list of factors emphasizes the meaningfulness of the context as it is only in this context that translations can be seen, analyzed and understood.

d. Room for contestation and for compliance

The levels (or domains) of policy are thus loosely connected, and an explicit focus on these connections suggests that there is equal room for contestation between domains as well as compliance among them. Explicitly analyzing connections is important as it allows to empirically establish how they happen, rather than merely assuming this. Scharpf (1978:347) emphasizes that "it is unlikely, if not impossible, that public policy of any significance could result from the choice process of any single unified actor. Policy formulation and policy implementation are inevitably the result of interactions among a plurality of separate actors with separate interests, goals and strategies". Therefore, the way and forms by which different organizations (actors) relate to each other becomes central to implementation studies (Hill and Hupe, 2002: 59).

Whereas early research of policy implementation either emphasizes the role of the policy designer or the policy implementer, such a clear distinction appears to have dissipated to the benefit of seeing both as an integral part of the process (Hupe and Sætren, 2015). From the second half of the 1980s onwards such 'synthesis' approaches gained traction (Sabatier, 1986). In these approaches the policy-making capacities of street-level bureaucrats are combined with the constraints posed by socio-economic conditions and legal instruments (Winter, 2012). In synthesizing the top-down and bottom-up approach, the synthesis approach also combines the different purposes of the analysis of policy implementation. Whereas the top-down approach focuses on the effectiveness of policy implementation and the bottom-uppers focus on how a complex network of actors shape policy, the synthesis approach combines these purposes. Although frequently championed as a bottom-upper, Lipsky's work hints at the influence of policymakers on the agency of implementers, bringing together these two domains in the policy process. While the concept of 'street-level bureaucrat' emphasizes the local agency and the importance of local adjustments to policy, Hill and Hupe (2002) suggest that the adjustments in which 'street-level bureaucrats' engage in is not necessarily a better interpretation of the needs on the ground but rather practices that reflect the inability of public officials to cope with the pressures they face (Hill and Hupe, 2002: 52). Actually, they may be developing "conceptions of their work and of their clients that narrow the gap between their personal and work limitations and the service ideal" (Lipsky, 1980: xii).

e. Organized hypocrisy

Brunsson's (1986) concept of organized hypocrisy can potentially offer body to describe the realities that implementers are indeed trapped by the 'freedom' of implementation of the ground, and the constraints these same implementers experience as they are part of a system. Organized hypocrisy is the result of the mechanisms organizations purposefully employ in order to gain support and legitimacy for their functioning from their institutional environment. Organizations, like public water utilities, which provide a basic service and operate in a complex institutional environment, depend for their survival on the support they receive from this institutional environment. In such a case, "organizational success depends on factors other than efficient coordination and control of productive activities. Independent of their productive efficiency, organizations which exist in highly elaborated institutional environments and succeed in becoming isomorphic with these environments gain the legitimacy and resources needed to survive" (Meyer and Rowan 1977: 352). In other words "organizations must demonstrate congruence with the values and norms of their environment in order to receive support" (Brunsson, 1986:165).

In addition to being congruent with the values and norms of their environment, these organizations also need to produce results for which the organization is established, and preferably do so in an effective and efficient manner. As highlighted earlier, water utilities are the main vehicle for ensuring access to water services of the population they serve. This service provision is a result these water utilities need to achieve. The result of actual water service provision, however, represents a different domain than the result of congruence with the institutional environment. Actual service provision is a result that needs to be achieved in the everyday domain. Congruence with the external environment is much more an element of the national and global policy domains.

In order to achieve these results organizations are able to employ three instruments. These three instruments are talk, decisions and actions (Brunsson, 1986). Although these instruments "are important in their own right" (Pollit, 2011:938), they are also intuitively interlinked. Talk, which Brunsson (1986:170) defines as "the spoken word", and decisions, which represents a type of talk that indicates a will to act, are "used for mobilizing and coordinating internal actions". For talk and decisions to be indicators for actions, it is important that talk, decisions and actions align. However, talk and decisions can also be externally-oriented rather than serving the purpose of guiding internal actions. "They are then used as ideological outputs of the organization, beside the output of products. By talking about themselves and others to external audiences, organizations are able to describe who they are and what their environment looks like, what and whom they like and dislike, what they try to do, what they actually do, why they succeed or fail" (Brunsson 1986:170-171). Externally-oriented talk and decisions may be captured in organizational policy documents or strategic plans. They may be presented at (inter)national conferences. They may be disseminated through articles, brochures and social media.

The idea of organized hypocrisy is that within an organization inconsistency exists between the talk, decisions and actions of the organization. Or as highlighted by Brunsson (1986) : "To talk is one thing; to decide is a second; to act is yet a third". Such inconsistency or hypocrisy is not necessarily to be understood in a negative way. Particularly for 'political organizations', which may face multiple conflicting demands and pressures such as achieving both social and commercial objectives of service provision, it may be useful and even rational to deviate the talk and decisions of the organization from the actions of the organization (Pollit, 2001). In other words, organizations may require to adhere to a particular talk and decisions in order to accommodate norms prevailing in a particular domain, but implement a diverging action in order to achieve an objective in a different domain. Organized hypocrisy is then viewed as a rational strategy that allows organizations to achieve multiple, to some extent conflicting, objectives. In this perspective the deviations between talk, decisions and actions that characterize organized hypocrisy "may even be a major promotor of success" (Brunsson, 1993:2).

4. Policy learning

Until here I have explored avenues of inquiry regarding how adjustments to the commonly understood commercial principles during implementation could be interpreted and analyzed. I have also explored the connections that link policy design to implementation and vice versa.

Puzzled by the lack of innovation in the current dominant models to providing water services, and guided by the hypothesis that there are adjustments currently taking place, in this section I explore avenues for understanding how these adjustments, could be informative for policy change.

Policies may be changed at different levels, and during different times in the policy process or cycle. They can be adjusted as they are implemented on the ground, at the lowest level they intend to influence. They can also be adjusted after a process of evaluation, in which the adjustments on the ground during implementation are identified and, if deemed appropriate, fed back into the policy formulation stage. This opens the possibility for learning. This learning would lead to change or adjustments. Such adjustments hint at the interlinkages between the different stages in the stages heuristic, as well as the non-linearity of the policy process. In this thesis policy change takes two meanings. First, change is understood as the adjustments done by implementers during the implementation stage, which has been discussed in the previous section. A second meaning of change is through policy learning. In this thesis this is defined as policy adjustment or change that takes place after implementation and through lessons learnt resulting from that implementation. There would appear to be ample room for such learning as problems that are unique to one country are usually not the norm[14] (Rose, 1991). Most issues that are of concern to the citizens of a particular location, such as education, social security, health care, economic development or water services, are shared across most locations. Given such similarities there would be room for learning from what has been done elsewhere.

Mollinga (2008:13) illustrates such "journeying […] in a bottom-up manner" by referring to the World Commission on Dams. "High levels of contestation among water user communities at the 'everyday' domain to state policies supporting dam construction led to the eventual development of a 'global' process to question policy assumptions. The World Commission on dams was an outcome of this process, and the report it developed in response has sought and iteratively aggregated the input of user communities for future policy development…". Relevant in Mollinga´s example of policy learning, is that it suggests that for a policy in the national policy domain (and even global domain) to be adjusted based on lessons learnt derived from policy implementation in the everyday domain, considerable feedback is required from a large number of everyday experiences, contestation and a considerable amount of time. Policy learning requires a "relatively enduring alteration of thought or behavioral intentions that are

---

[14] Rose (1991) provides the example of the reunification of Germany as policy issue that is unique to a particular location.

concerned with the attainment (or revision) of the precepts of a policy belief system" (Sabatier, 1993: 19). As such, policy learning implies "adjusting understandings and beliefs related to public policy" (Duplop and Radaelli, 2013 in Moyson et al, 2017: 162). Different processes may lead to policy learning such as changes due to success or failure of policy implementation (Hall, 1993), adjustment as a reaction to the environment (Heclo, 1974), or learning derived from lessons drawn from and by other governments in other regions (Rose, 1991). Whatever the underlying influence may be, policy learning essentially revolves around three questions: who learns, what is learnt and what effect does the learning have in the changing of policies (Bennet and Howlett, 1992: 278)? Important in this perspective is also that policy learning is very much a political process. Changes in policy impact the interests of actors.

In understanding the political nature of policy learning it is also important to examine the type of learning that is being done. Some learning may be more or less political than other types of learning. Broadly speaking, policy learning can be categorized in three types of learning (Sabatier and Jenkins-Smith, 1993: 221):

— *Instrumental learning*: This type of change will manifest itself in the instruments used to achieve a goal. This learning would occur when aspects of the problem are commonly seen as serious, causal links between the identified problem and the objective can no longer be laid, or the appropriateness of government instruments such as administrative rules, policies or funding arrangements are questioned. These learnings are moderately easy to realize in policy changes as they require new 'means' to be devised, while the overarching policy objectives do not have to change. This type of learning would lead to what has been referred as first and second order changes (Campbell, 2002: 23). A first order change modifies a policy instrument slightly (and in small increments), without questioning the principle of the instrument. A second order change implies more visible changes to the policy instrument, but it is still informed by practical issues during the implementation of such instruments (Cairney, 2011). An example of policy learning in this order concerns the experiences with the implementation of increased block tariffs (IBT) by water utilities. In theory, the block tariff of the IBT places consumers that use a lot of water in a higher block tariff and thus forces them to pay more for their water. The higher tariff paid by such consumers, which would be assumed to be high-income consumers, can then be used to cross-subsidize lower-income consumers, who are thought to consume relatively little water. The suggested benefits of IBTs, however, did not materialize during implementation (Whittington, 1992). In its actual implementation such cross-subsidization hardly materialized as 1) the proportion of users in higher segments is too low to cover for the proportionally bigger lower consumption segments, 2) low-income consumers that share a connection end up in higher tariff blocks, 3) the higher tariff blocks are too low to really allow for cross-subsidization and 4) those that would benefit most from a subsidized water tariff are not connected to the network to begin with. Water utilities, in an effort to correct this, have slightly changed tariffs per each block, or have slightly modified the minimum rate for a minimum consumption. However, the underlying policy principle of payment for water and the concept of water as an economic good remain untouched. In other words it is the policy instrument that is tinkered with rather than the policy principles.

— *Conceptual or problem learning* requires seeing things from a different evaluative viewpoint. These changes require the adoption of new concepts, principles and language to describe the problem (Kemp and Weehuizen, 2005). These learnings will manifest themselves in a fundamental change of the policy position, including its core values and strategies to achieve it. This type of learning is more difficult to occur than instrumental learning. Often, for this learning to lead to an actual policy change an external shock is required. As long as this shock does not take place the core beliefs of a policy will remain intact (Sabatier, 1993: 34)[15].

— *Social learning*: implies learning about values and other "higher-order-properties such as norms, responsibilities, goals, and a new framing of cause and effect relationships. Changes from this type of learning are rare as they imply an almost 'religious conversion' touching upon the nature of man, relative priority of values such as freedom, security, health or knowledge" (Kemp and Weehuizen, 2005: 11). Social learning and conceptual learning would translate into a third order change (Campbell, 2002: 23). A third order change is rare and it can be compared to a paradigm shift (Kuhns, 1970 in Cairney, 2011). "Policymakers customarily work within a framework of ideas and standards that specifies not only the goals of policy and the kind of instrument that can be used to attain them, but also the very nature of the problems they are meant to be addressing (…) this framework is embedded in the very terminology through which policymakers communicate about their work, and it is influential precisely because so much of it is taken for granted and unamenable to scrutiny" (Hall, 1993: 279). Hall (1993) argues that for this paradigm shift to occur policy failure has to

---

[15] See 'selective perception and partisan analysis' (Sabatier, 1993:34)

be of such proportion that also a shift in power at the policymaker or policy proponent level needs to occur as well, displacing the people and the ideas with them.

These three types of learning presented above suggest that learning is not only a matter of capacity to learn, but also – and perhaps of greater importance – is dependent on the political willingness and momentum to adjust. As such, learning can be 'random, biased or even absent altogether' (Dussauge-Laguna, 2012; Shipan and Volden, 2012; Wolman and Page, 2002 in Moyson et al., 2017). These approaches also acknowledge that policy learning, as policy making and formulation, is not a fruit of individual endeavors. Policy happens in and through many layers of society representing various interest groups. These layers permeate both government and society (Sabatier and Jenkins-Smith, 1993). As such, the explanations to identify whether learning occurs may not exclusively reside at the policymaking level, but also in the way information travels between these layers. Second, decisions and policymaking are bounded by the environmental and social constraints in which the learning occurs (Cairney, 2011). This implies that policy learning needs to be seen in the broader context in which it occurs, and what contextual factors foster or inhibit the processing of 'lessons'. Third, these approaches understand policy making as a historical process and a product of time (Sabatier, 1993). Policies are never started from scratch. Rather, they already carry a history with them, be it in the ideas on which they are underpinned or the instruments that have been used in that specific setting.

Methodologically this poses a challenge, because it makes the concept of learning slippery. Argyris (1976) argued that learnings cannot be pinpointed to specific behaviors as beliefs and actions related to such learning can manifest themselves in a very different, and even contradictory, manner. This inconsistency may not only be difficult to trace, but also even deliberately produced by the actors involved (Easterby-Smith, 1997).

Chapter 4: Methodology and Research Design

In this chapter I explain how I operationalize the theories and concepts described in the previous chapters to guide the collection of data in order to answer my research questions. As explained, I wanted to focus on the doings and daily adaptations that operators and other relevant agencies engage in giving meaning to a policy model during implementation assuming that the policy implementation is one of necessary translation and modification. For the operationalization of the data collection, I introduce the concept of 'service modalities' and a framework for analyzing the implementation of commercialization in order to describe and compare how utilities in different contexts deal with and adopt the policy principles of commercialization that they are asked to follow. Later in the chapter I elaborate on the justification of the case studies and the focus on small towns for this research. I argue that small towns offer a particular set of characteristics that pose very clear challenges to the implementation of commercial principles.

1. Studying adjustments

Mukhtarov (2014) observed that one of the main challenges of analyzing policy transfer as a process of translation, consisting of adjustments to specific contexts, is that the policy (model) is often not clearly delineated and is therefore difficult to neatly identify and subsequently compare. To operationalize a study of policy translation, therefore, it is pertinent to find a way to characterize the policy model that is studied. This characterization sets the boundaries of what is studied. In chapter 2, I defined the model of commercialization to consist of two principles. A commercial water provider should operate as an autonomous entity and operate on the basis of full cost recovery.

The second challenge is to develop a framework to analyze how these principles are implemented in the different case study locations. This framework consists of two parts. First, the cases will be described in terms of a service modality consisting of a combination of infrastructural and technological, organizational and financial arrangements shaping the provision of water services. Second, the implementation of the principles of commercialization will be analyzed using a framework originally developed for the study of concession contracts as it helps identify major factors that justify the choices made in operations.

a. Service modalities

Water service provisioning models encompass a particular configuration of infrastructure, the organization that operates and manages that infrastructure and an arrangement to (re-)cover the cost of providing services.

The infrastructure used to provide services may range from a sophisticated centralized network, which is usually deemed to require relatively advanced capacity and skills to build, manage and operate, to decentralized technologies such as hand pumps or standpipes, which are deemed to be simpler to build, operate and maintain. Different infrastructures are usually associated with their own, more 'appropriate', type of organization or management structure, often in relation to their perceived level of 'technical sophistication'. Hence, sophisticated large-scale centralized water systems are usually operated by formal (large) water utilities, which are expected to have the necessary technical, commercial and financial capacities to run such a system. Within these organizations, a considerable degree of technical and managerial specialization of different staff members is expected. Decentralized water systems are usually characterized by smaller organizational setups, in which a larger role may or can be attributed to users who do not necessarily have much specialized knowledge or skills (Solo et al. 1993:6). Finally, service models can be described by their financial regime, i.e. the ways in which costs involved in providing services are to be (re-)covered. Financial regimes do not only differ in terms of their sources of funding (tariffs, government subsidies, etc.), but also vary depending on the technology used, the type of organization and the market that is being served as these determine both how much funding is required as well as the potential to (re-)cover costs.

Figure 2 Elements of water service provision modalities (Source: WSM Group IHE Delft, 2017)

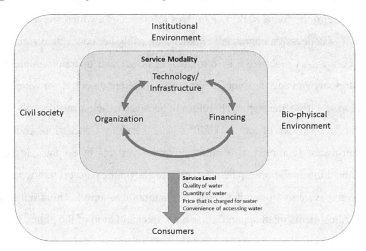

Different configurations of infrastructure, organizations and financing arrangements lead to different levels of service for consumers. A service level can be characterized by the quality and quantity of water consumers can access, the price that is charged for water and the convenience (distance, time investment) of accessing water.

In addition to the three core elements of every water service delivery model, it is important to note that the provision of water services happens in particular socio-economic, political and bio-physical environments that influence the delivery of services. Hence, the availability and quality of raw water resources, the topography, socio-demographic developments, legal and administrative frameworks and culture, and 'civil-public' interactions (Larbi 1999; Schwartz, 2004) all co-shape how services are and can be provided, as well as resulting service levels. Following Mollinga's conceptualization of water resources management, the elements influencing practices of water services provisioning can be categorized in three dimensions: "the ecology and physical environment, the ensemble of economic relations, and the institutional arrangements of state and civil society" (Mollinga, 2008: 11).

b. Framework for analysing the implementation of commercialization

The description of service modalities is the first step in the operationalization of my study. A second step in this operationalization is to find a way to analyze how the prevailing principles of commercialization shape and influence what utilities and operators actually do and why? In order to do that I make use of the framework by Komives (1998; 2000), which she developed for analyzing the design of concession contracts[16]. Rather than using her conceptualization for designing concession contracts, I mobilize it here to analyze how water operators implement commercialization (autonomy and cost-recovery). Despite the different objective for which this framework was developed I find the approach relevant because it helps me describe two important and relevant manifestations of commercialized water services. Concessions contracts are based on both ensuring the (commercial) viability of water systems by emphasizing cost recovery as well as establishing a relationship between private and public partners that separates the operations of the water systems from other (political) actors (autonomy). This framework allows me to trace the adjustments that happen during the implementation of the policy model and it also helps in identifying the motives and motivations of operators for making these adjustments.

In the framework for analyzing the implementation of commercialization, based on Komives (1998; 2001), I identify a number of relevant elements:

1. Revenue generation: This relates to the income generated by the water provider. Most often three different funding sources for water providers are distinguished. These are frequently referred to as the 3T's (OECD, 2009; World Bank, 2017): tariffs, taxes and transfers. Elements that are important include sources of revenue and strategies for increasing revenue generation. A specific example of such an element is the degree to which a water provider is able to set the tariffs they charge consumers.

2. Costs: In providing water services a water operator incurs different types of costs. The most common distinction between costs relate to capital expenditure (CapEx) and operational expenditure (OpEx). Although variations and more detailed breakdowns of these types of costs exist (see Franceys et al, 2016), I distinguish capital expenditure (CapEx) and Operational Expenditure (OpEx).

3. Time of engagement: This time refers to the period in which the water operator needs to balance the revenues and costs for providing services. In case time is relatively

---

[16] In water services literature, five types of contracts for private sector participation are usually distinguished of which concession contracts are one (World Bank, 1997). In a concession contract the private operator takes over the delivery of infrastructure services in a specified area for a specified time (usually 20-30 years), including all related operation, maintenance, collection and management activities. Moreover, the concessionaire is responsible for any capital investments in the system. The concessionaire is financed by the revenues that it is able to collect from the consumers.

limited, the operator either needs to generate substantial revenues (increased tariffs, large subsidies, etc.) or significantly reduce costs of operation and investment. If the operator has more time to balance revenues and costs, it can spread out the recovery of costs over a larger time period (in principle either allowing for lower tariffs and/or more investments in the system).

4. Market characteristics and size: The balancing of revenues and costs happens in relation to a particular configuration of consumers that the operator serves. Important dimensions of the market composition include elements such as the size of the market in terms of the number of consumers and the quantities of water consumed, the income level of consumers, the population density, the degree to which the water operator competes with other water providers, and the availability of alternative sources of water.

5. Biophysical and Governance Context: The four previous dimensions interact to reveal the way in which commercialization is implemented. However, these dimensions do not operate in a vacuum, but are strongly impacted by the bio-physical and governance context in which the water provider operates. As highlighted in the section on service modalities, the biophysical context impacts costs of providing services, but also the degree to which consumers have access to alternative sources (market conditions). The governance context consists of a policy and legal framework that regulates the operations of the water provider, the institutional landscape in which the provider operates (government agencies, donors, etc.) and the relationship of the provider with civil society and the consumer. The governance context greatly impacts the ability of the provider to generate revenues and influences the type and level of costs that the provider will have to incur.

Figure 3 Factors to study in implementation of commercialization principles in water services (Source: Based on Komives, 1998; 2001)

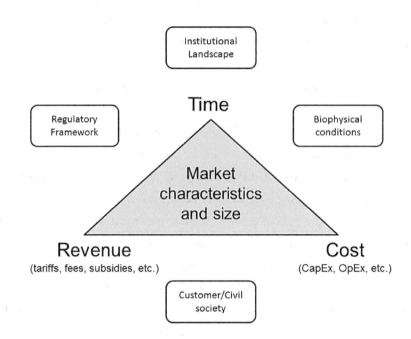

2. Case study selection: The challenge of commercialization in small towns

> 'A first class water works man in a small town needs the wisdom of a Solomon, the patience of a Job, the political tack of a Jim Farley, the tenacity of a General McArthur, the diplomatic ruthlessness of a Von Papen, the strength of a Pittsburgh steel worker and a doctor's degree in engineering' Dillery, 1945: 118).

I find this quote of Dillery (1945) particularly illustrative of the operational challenges of managing a water system in a small town and that helps me justify the particular focus of cases chosen for this thesis. Earlier I have questioned the validity or usefulness of commercial principles for the sector, and most practically, for the operators themselves. However, small towns have specific characteristics that could particularly challenge these principles. First, small towns are characterized by very low population densities which hamper the realization of economies of density that can be realized in more densely populated areas in large cities. As a result, centralized systems may not represent a (financially) viable service modality as the costs of such infrastructure cannot be recovered from the revenue that the provider can obtain from the users constituting its market (Rondinelli, 1983; Adank, 2013). Secondly, in small towns an apparent mismatch exists between the technical and financial capacity at the local level and the requirements demanded by increasingly complex water supply systems (Mugabe and Njiru, 2006; Mara and Alabaster, 2006). Many small towns simply do not have sufficient capacity to operate and maintain relatively complex water systems. Thirdly, the dynamic and relatively fast growth of small towns often requires the implementation of more flexible water delivery systems. Most approaches to water services delivery, however, are fixated on fixed designs of treatment plants and distribution networks (Lauria, 2003).

What makes the development of water infrastructure more challenging is the lack of information on current developments in small towns as well as poor planning strategies (Moriarty et al, 2012; World Bank, 2017; WaterAid, 2010). As a result, the access to water services in small towns in typically lower than in other settlements. A study of the US National Academy of Sciences based on data from more than 90 countries showed that residents of smaller cities[17] suffered a clear disadvantage in the provision of basic public services (i.e. piped water, waste disposal, electricity and schools) than residents of larger urban centers (Cohen, 2006)[18]. The challenges of service provision in small towns can be summarized in three main dimensions (Moriarty et al., 2012; Adank, 2013; Mugabi and Njiru, 2006; Mara, 2006; World Bank, 2017):

— High cost vs. low revenue: The distribution of people in small towns is generally less dense than in bigger urban areas. This makes it difficult to reap economies of density and economies of scale in infrastructural development.

— Existing vs. required capacity: The technical and financial capacity at the local level is usually limited.

— Flexibility vs. rigidity: The dynamic and relatively fast growth of small towns often requires the implementation of more flexible structures rather than established designs of treatment plants and distribution networks.

---

[17] Note that the smaller category according to this study is 100.000 inhabitants. Even though there is a mismatch in terms of population numbers in other studies used in this discussion, we could expect these trend to reflect equally disadvantageous situations in smaller urban centers.
[18] No specific data on small towns at aggregated level is available since 2006 (Kingdom, 2005, UN-HABITAT, 2006, Cohen 2006)

It is noteworthy that these challenges are not restricted to developing countries. A recent review of literature on 'small drinking water system governance' in industrialized countries reveals similar challenges (McFarlane and Harris, 2018). Systems in small communities in either semi-rural or peri-urban areas in industrialized countries suffer equally from a small customer base and lack of financial support, as well as limited local capacity or operational and managerial support to comply with regulatory requirements relating to service levels and, more importantly, public health standards (McFarlane and Harris, 2018). The relaxing of standards imposed on these systems to reduce the operational and financial burden has been criticized for creating a two tier protection level for different users depending their point of access (Daniels et al., 2008 in McFarlane and Harris, 2018).

Practitioners and policy makers have little apparent information or knowledge on approaches to service provision in small towns[19]. Instead, what we know about development of water services in small towns is understood through either 'big theories' on urban development or through well-documented approaches to rural water services. Settlements are then qualified as being either urban or rural (UN-HABITAT, 2006). However, these urban or rural approaches lack reflection on the actual characteristics and complexities of small towns, making the application of these approaches to small urban centers problematic, if not impossible (Bell and Jayne, 2009: 684). When research is specifically developed to unravel processes of the governing of small drinking water systems they are usually focused on single governance dimensions such as regulation or finance (MacFarlane and Harris, 2018). The result is that standardized models for water service provision are implemented in these areas where perhaps other approaches are required (Choquill, 1989).

The choice to study the implementation of water services models specifically in small towns is underpinned by the conviction that most of these models, originally envisioned for either traditionally urban or traditionally rural settings (Pilgrim et al., 2001), require considerable adaptation to respond to and cater for the specific contexts of small towns, which are located between the typically urban and the typically rural and may therefore require a different approach altogether (Hopkins, 2003). The delegated management and commercial utility model were initially developed for large urban centers. Their implementation in the context of small

---

[19] Interestingly, one of the few streams of literature on water services in small towns concerns the challenges described of servicing small towns in the United States. This literature from the middle of the 20th century is largely collected in engineering journals (highlighting that water services provision in that era was still mainly viewed as an engineering challenge) (Dillery, 1945). It is only at the end of the 80s that the complexities of servicing small towns go beyond an engineering approach.

towns, which exhibit both urban and rural characteristics, is likely to demand considerable modifications. Community Based Organizations, which are usually promoted for the expansion of services in settlements with rural characteristics, may also require adaptations in settings that are currently rapidly transforming themselves to more urban-like settlements. This makes small towns an interesting research location for analyzing particular adjustments of these service provision models in practice.

3. *Defining small towns*

In urban studies considerable efforts have been made to define 'the urban' (Robinson, 2005). Once areas are classified as urban, an automatic classification of 'the rural' follows by default. Essentially, everything that is not urban is considered rural. How small towns fit in this urban/rural dichotomy is not entirely clear. One of the challenges of defining small towns is that it is virtually impossible to find fitting criteria to define such settlements. In the provision of water services, the neat clustering of infrastructural and organizational models along an urban-rural classification is similarly problematic. This in-between nature of small towns is also apparent in how Pilgrim et al. (2004) define small towns. They suggest that these settlements are "sufficiently large and dense to benefit from the economies of scale offered by piped systems, but too small and dispersed to be efficiently managed by a conventional urban water utility" (Pilgrim et al., 2004). These small urban centers will typically appear to have rural socio-economic contexts while requiring urban-type technology for water provision (Hopkins, 2003).

For some, the dichotomy of urban versus rural is increasingly losing its relevance (Noronha and Vaz, 2014). However the labels of urban and rural still have a significant value. These labels provide a particular understanding of how the settlement is placed in the administrative government system. This placement determines the allocation of budgets and assistance programs, and the degree of autonomy to execute projects. As such, this categorization determines what resources a settlement has access to. Labelling settlements "is the counterpart of access in that authors of labels, of designations, have determined the rules of access to particular resources and privileges. They are setting the rules for inclusion and exclusion, determining eligibility, defining qualifications" (Wood, 1985:352). Moving from one category to another then also allows settlements to gain access to certain resources that otherwise would not be within their reach. As Samanta (2014:56) explains about the 'census towns' in India, the simple categorization of a settlement as panchayats (in specific states) irrespective of their size or economic activity, determines their capacity to raise, collect and administer taxes and therefore, eventually also provide better local services.

Rather than present a singular definition of what constitutes a small town, this research has adopted the working definitions of small towns in the three countries studied. Two reasons underlie this choice:

1. Definitions of small towns differ per country and are subject to the relative size of national populations, economic geographies and its related social distribution in a space or territory (Roberts, 2016). Similarly, finding a universal definition that would allow urbanization phenomena in different countries to discern global standards of 'small towns' also poses challenges. As such, choosing a figure or certain phenomena as being representative for small towns is not possible.

2. Even though hard to define, definitions in this research play an important role as the labels of urban and rural do have a value. Administrative boundaries and "the rationale for defining these boundaries are accompanied by political priorities and therefore have compelling implications for policy development" (Hopkins and Sattherwaite, 2003:2).

4. Selected Cases

Given the focus of researching the practical application and adaptation of models for the delivery of water services, the research follows a case study approach (Yin, 2012). This case study research focuses on each of the three dominant models. The selection of case studies is as follows:

1. **Commercial Public Utility** – This modality is researched through the case of National Water and Sewerage Corporation (NWSC) of Uganda. The reasons for selecting NWSC as a case study are:

    a. The utility has consistently presented itself as a successful example of a commercial public utility. NWSC has presented itself as a 'turn-around' utility that combines public sector control with private sector efficiency (Muhairwe, 2009). The NWSC has also received international recognition as such. NWSC was recognized as the African Utility of the Year from 2014-2016 during the annual African utility week. In 2016, the water operator was presented as "one of Africa's model utilities providing new performance frontiers and most importantly supporting the performance improvement in other utilities across Africa and beyond" when it was voted African Utility of the Year for a third consecutive year[20].

b. IHE Delft has a Memorandum of Understanding and a good working relationship with NWSC allowing for access to key-informants and information. Also, IHE Delft and NWSC are both involved in the SMALL[21] (DUPC2) and the ATWATSAN[22] (FDW) project, which allows for research activities as part of this PhD to be funded and for access to information to develop the case study.

c. Whereas NWSC was previously responsible for servicing the large towns in Uganda, its mandate has expanded since 2013 to also service small towns. This has meant that the number of towns serviced by NWSC increased from 23 in 2010 to over 250 in 2018. Despite the challenges that appear in servicing these small towns the NWSC continues to represent itself as a commercial public utility.

d. The National Water and Sewerage Corporation has received considerable attention from researchers analyzing its operations and performance. As such, considerable secondary data is available for the case of NWSC.

In Uganda, two service areas were analyzed. The towns of Bushenyi en Ishaka form the main small towns in the Bushenyi/Ishaka service cluster. Bushenyi has an estimated 42,000 inhabitants and is located in the south-western region of Uganda. The customer base in the Bushenyi/Ishaka cluster counts approximately 9,600 connections. Through these connections approximately 65% of the population in the service area is served with potable water (NWSC Bushenyi, 2017). The second service area that was analyzed is the small town of Kitgum. It has a population of about 44,000 people, which are serviced through approximately 3,000 connections. The service coverage is estimated at 80% in Kitgum (NWSC Kitgum, 2017). Whereas Bushenyi has been part of NWSC's mandate since 2002, Kitgum was only transferred to the NWSC in 2013 in the first wave of small towns that were transferred under the new Government policy to allocate small towns to NWSC.

---

[20]http://www.monitor.co.ug/News/National/NWSC-voted-African-utility-of-the-year-2016/688334-3215614-tx7qx1z/index.html

[21] https://small.un-ihe.org/home

[22] Alternative Approaches and Tools for Improved Water Supply and Sanitation for Towns in Northern Uganda

Table 3 Drinking water provision in small towns in Uganda

Element	Indicator	National Water and Sewerage Corporation
Commercialization	Legal Status (autonomy)	Parastatal – meaning the utility is a government organization operating under public law, but has a separate legal entity through Decree 34 from 1972.
	Regulation (autonomy)	The NWSC Head Office engages in a three-yearly performance contract with the Ministry of Water and Environment
Finance/Commercialization	Cost recovery	The 1999 National Water Policy dictates that "financial viability of public utilities should be assured". In utility-operated urban schemes tariffs should "cover repayment of construction loans, depreciation of technical installations (i.e. replacement costs), and full cost of operation and maintenance" (Ministry of Water Lands and Environment, 1999: 18). The NWSC's Five Year Strategic Direction pinpoints financial sustainability as one of its four Strategic Priority Areas
	Capital investment	Although the water utility should strive for full cost recovery, the National Water Policy highlights that "the Government will continue to offer subsidy to the majority of water supplies until adequate financial and management capacities are developed at the districts and urban councils" (Ministry of Water Lands and Environment, 1999: 18).
Infrastructure	Ownership of infrastructure	All assets are owned by the NWSC Head Office
	Type of Infrastructure	The 1999 National Water Policy highlights that "appropriate low-cost technologies should be selected" (Ministry of Water Lands and Environment, 1999: 17).
		The Five Year Strategic Direction suggests an approach of service differentiation as the utility states that "public standpipes remain a major means of providing services to the urban poor in informal settlements" (p.14). Moreover the Direction also highlights the systematic increase of pre-paid public water points.
Management	Operation of infrastructure	NWSC branch or area offices are responsible for operational activities. They hold a performance contract with the NWSC Head Office to monitor their performance.
Universal Service Coverage	Universal Service Coverage	The National Development Plan II (2015-2020) dictates that NWSC is expected to increase water coverage to 100% by 2020. The NWSC Five Year Strategic Direction sets the target of achieving 100% coverage by 2021.

2. **Delegated Management** – This modality is researched through the case of the private operators providing services under the Administration for Water and Sanitation Infrastructure (AIAS) in Mozambique. AIAS, which acts as an asset holding company, was founded to exclusively develop water services in small towns in Mozambique in 2009. The reasons for selecting this particular case are as follows:

   a. During my MSc research I studied small-scale private operators in the city of Maputo. During this research my interest in small-scale private operators in small towns was awakened. From my MSc research I had also developed knowledge and a network of key informants that were of great use in setting up this research and facilitated the collection of data in Mozambique.

   b. IHE Delft is involved in close cooperation with Vitens Evides International (VEI) in a number of different projects. VEI is a Dutch international water operator that develops international non-profit projects for Dutch water utilities. VEI is closely involved in the institutional development of AIAS through a technical assistance contract. Through this cooperation the initial contacts within AIAS, and its operators, have been established.

   c. AIAS and VEI are both partners in the SMALL project. This project provides funds to support activities related to this PhD research in Mozambique. Moreover, this partnership also allows access to key-informants and information relevant for developing this case study.

In Mozambique this study focused on the operations of three different small-scale private operators in the towns of Moamba, Caia and Manjacaze. The operator in Moamba currently operates 9 other towns and the operator in Manjacaze another additional 2 towns. Moamba was the first system to be tendered in 2013. It has 5,000[23] inhabitants and in 2013 approximately 50% of the population was connected to the system through household connections or in-yard connections. The other systems managed by the same operator vary in size between 8,000 inhabitants (Homoine) to 43,000 inhabitants (Mopeia). The second operator manages three systems under AIAS contracts. Manjacaze was only signed in 2016. At that time this town had about 10,000 inhabitations and only 8% of the population had access to water. The other two systems are located in Nametil (65,000 inhabitants and 11% coverage), and one in Espungabera (6,000 inhabitants and 79% coverage). The third operator runs the system in Caia. It is the only system this operator runs under an AIAS contract. When the contract for Caia was tendered the

---

[23] All population figures are projected from the last census carried out in 2007

population was estimated to be about 25,000 inhabitants with a service coverage of about 15%. In addition to these towns, other systems such as Mopeia, Bilene or Namaacha where researched, either through direct visits or through interviews with the operators serving these towns.

3. **Community Management/DRA** – The research of the community management/DRA model is developed by focusing on eight community based service providers (CBO) in Lamongan Regency in Indonesia. These CBOs were the 'Tlanak' Village HIPPAM, 'Kemlagi Gede' Village HIPPAM, 'Geger' Village HIPPAM, 'Karangwedoro' Village HIPPAM, 'Sukomulyo' Village HIPPAM, 'Trepan' Village HIPPAM, 'Pawer Siwa' HIPPAM, and 'Pengumbulanadi' Village HIPPAM. In these cases, the initiative to establish the CBOs lay with the World Bank.

Located in the eastern part of Java, Lamongan Regency is a rapidly growing municipal district due to its proximity to the capital city of the East Java, Surabaya. By last available official data the population of Lamongan Regency[24] is estimated to be 1,187,084 people[25]. The regency is divided in 27 sub-districts and 262 villages[26]. Water services in Lamongan Regency are provided by different service providers in the area. The water utility (PDAM), owned by the municipal government, provides services to the urbanized area of the Regency and supplies about 67% of the inhabitants with water (BPPSPAM27, 2014). The selection of these CBOs in Lamongan Regency is based on three criteria:

1) The CBOs had run with profits since their establishment
2) The CBOs had been able to access funds from commercial entities; and
3) The CBOs had expanded their network/customer base to different degrees.

The selection of CBOs based on these criteria was informed by the focus of the original MSc research of Riski Aditya Surya (IHE Delft 2015/2017), who was interested in understanding why these CBOs in Lamongan region had been performing relatively well whilst in other parts of Indonesia CBOs struggled to sustain themselves.

---

[24] In the Indonesian administrative system a distinction is made between cities and regencies. Regencies are characterized by scattered rural areas within its administrative boundaries.
[25] Badan Pusat Statistik Kabupaten Lamongan – Lamongan Statistic Bureau. www.lamongankab.bps.go.id/linkTabelstatis/view/id/412 - retrieved 20 September 2016
[26] idem
[27] BPPSPAM is National Supporting Agency for Water Supply System Development which operates under the Ministry of Public Works and Housing

Table 4 Drinking water provision in small towns in Mozambique

Element	Indicator	
Autonomy	Legal Status (autonomy)	"[AIAS] is created as a parastatal organizations with administrative autonomy" (Decree 19/2009 art. 1).
	Regulation (autonomy)	"The water systems in cities and small towns need to be managed by autonomous organizations operating on the basis of commercial principles" (Politica de Aguas, 2016 – art. 2.2).
Finance	Cost recovery	"Besides its social and environmental value, water is also an economic good. Water is important for the economic development and reduction of poverty. To allow water services to be financially viable the price of water should be close to its economic value" (Politica de Aguas, 2016 – art. 1.3.c). "AIAS will promote that operations are autonomous, efficient and financially viable through the delegation of services to private operators or other organizations" (Decree 19/2009 – art. 2).
	Capital investment	It is responsibility of AIAS to support "the capital costs associated to exploitation and maintenance of assets, equipment and services necessary to provide water services" (Resolution n. 34/2009 – art. 18). "The Government (of Mozambique) will provide the main source of funding for the development and rehabilitation of water infrastructure" (Politica de Aguas, 2016 – art.2.2).
Infrastructure	Ownership of infrastructure	AIAS is the owner of the assets
	Type of Infrastructure	"The infrastructure for rural areas and small human settlements are wells and boreholes with handpumps, rainwater harvesting or protected wells. The Government (of Mozambique) prioritizes the development of small water systems (*conventional system*) to small towns depending on their development" (Politica de Aguas, 2016 – art. 2.1).
Management	Operation of infrastructure	The daily operations are delegated to a private operator holding a contract for five years with revision of two more.
Universal Service Coverage	Universal Service Coverage	The Government of Mozambique endorses SDG 6 in its latest Water Policy, which was revised in 2016 (Politica de Aguas, 2016).

Table 5 Drinking water provision for rural water in Indonesia

Element	Indicator	
Organization	Legal Status (autonomy)	National Policy on Community Based Water Supply and Sanitation Services which allows for delegated management to a *Himpunan Penduduk Pemakai Air Minum (Community Association of Drinking Water Users)* (National Planning Agency, 2003). HIPPAM in Lamongan Regency were established on the basis of the Governor Regulation No. 11 Year 1985 "HIPPAM Establishment".
	Regulation (autonomy)	HIPPAMs are regulated as part of individual statuses of establishment defined by each community but are not accountable to regional or national regulations.
Finance	Cost recovery	Donor-funded projects (WSLIC and PAMSIMAS) instilled a focus on 'sustainability', which they linked to the concepts of 'full cost recovery', 'demand responsive approach' and 'sanction and disconnection'. This focus was to "bring CBOs into more commercially viable community enterprises as well as in reducing the financial burden on generally financially-strapped local governments" (WSP, 2011).
	Capital investment	The donor-funded projects funded the development of initial infrastructure to exploit water resources. Upon establishment it is up to the HIPPAM to operate and maintain, and were applicable, expand the infrastructure
Infrastructure	Ownership of infrastructure	The infrastructure is owned by the community, as represented by the HIPPAM.
	Type of Infrastructure	"simple low-cost technology" (PAMSIMAS Project document)
Management	Operation of infrastructure	The HIPPAM is responsible for the operation and maintenance of infrastructure.
Universal Service Coverage	Universal Service Coverage	The Indonesian Government aims to ensure universal access to sanitation and drinking water by 2019 – commonly referred as the '100-0-100 targets' (universal access to water, eliminate all slum areas, 100% access to sanitation facilities)

5. Data collection

In developing these case studies, data was collected in a number of different ways. Data collection has been developed by myself, as well as through MSc students that I mentored at IHE Delft. For Mozambique, the data used to develop this case study was collected during fieldwork activities from June 2015 to February 2018 in five different episodes of data collection. The data has been collected through 46 semi-structured interviews with relevant stakeholders. Interviewed respondents included representatives or consultants of the implementing agency (AIAS) in the sector, representatives of Directorate of Water (DNA), the regulator CRA, technical advisors and small-scale private operators. The data analyzed is derived from different systems. For some of them (Moamba, Caia, Bilene, Namaacha and Manjacaze) local authorities such as the Town Council or the regional representation of the Ministry of Public Works and Water Resources were also interviewed to understand their position concerning the involvement of the private sector in the provision of services. Official documents such as contracts, regulations, and consultancy reports have been used to complement the data collection process. Additionally, performance data in relation to contractual KPIs have been accessed through the involvement of two experts in the actual implementation of contracts with private operators in small towns in Mozambique.

For Indonesia, data collection was conducted from November 2016 to January 2017. The data collection for this research was based on 42 semi-structured interviews. Eighteen interviews were carried out with CBOs as they were the main focus of this research. During these interviews the financial statements of the CBOs were also examined to gain a better understanding of the financial performance of the CBOs. In addition, and to provide necessary background and context to the development of CBOs in the Lamongan Regency, another twenty-three interviews were carried out: CBO Association (3), consumers (8), Village government representatives (2), Local government agencies (2), provincial government agencies (2), Ministry of Public Works and Housing (1), international donor agencies (3) and Indonesian water sector experts (2). Data collection was designed and undertaken by Risky Aditya Surya as part of his MSc thesis research at IHE Delft (2015/2017). Mr. Surya was at the time of the research Advisor to the Indonesian Agency for Support to Water Operators (BPPSPAM). I was the MSc. mentor of Mr. Surya during his research at IHE Delft. In addition, under the NUFFIC project IDN/186 the Ministry of Public Works of Indonesia organized a training for capacity development of the Heads of CBOs in Bekasi, Indonesia in October 2017. I facilitated several sessions during this workshop.

In Uganda, research data was mainly collected from November 2017 to January 2018 by Maxi Omuut. Respondents for semi-structured interviews were selected based on their positions and direct involvement in planning, designing, operationalizing and management of infrastructure and finances of water supply services at NWSC. The respondents from the NWSC Head Office included a director, senior managers and engineers. At the NWSC operational towns of Bushenyi and Kitgum, the respondents included the area managers, the heads of the technical, finance and commercial sections. In addition, some technical staff were also interviewed. In total, 22 respondents were interviewed. Moreover, a focus group discussion was held at the Head Office of NWSC in March 2018 with a representation of 35 Area/Branch Managers. In addition, secondary data like monthly operational reports, financial statements, annual CapEx and OpEx budgets, bi-annual performance evaluation reports and audited annual reports were reviewed for the period 2013 to 2017 to determine the performance of the towns. Furthermore, reviews were undertaken of NWSC documents relating to current development programmes, policies and action plans on infrastructure, and national water supply and sanitation standard designs. The Government of Uganda performance contract with NWSC was also analyzed as this document also provides insight into the state of affairs at NWSC Head Office and service delivery in the towns. In addition, the project ATWATSAN funded a workshop in view of the upcoming contract review between Head Office and Areas/Branches. I led the facilitation of this workshop in February 2018. This research has mainly been developed under the framework of the project SMALL (Water and Sanitation in Small Towns – DUPC2) by Maxi Julius Omuut for his completion of his MSc research at IHE Delft (2016/2018). During the implementation of the research, Mr. Omuut was employed by NWSC. During the MSc research of Maxi Julius Omuut, I was his MSc mentor and, as such, closely involved in his research from the development of the research proposal, data collection to analysis.

Chapter 5: Delegated Management Framework in Small Towns in Mozambique[28]

In this chapter, I document and analyze how the paradigm of the commercialization of water provisioning took shape and was implemented in small towns in Mozambique. The overall policy model that was adopted in Mozambique for the provision of water services is that of the Delegated Management Framework (DMF). This is a model that proposes a differentiation between the asset holder or the owner of the infrastructure forming the water systems and the operator(s) of these systems. The idea is that assets for providing water services remain public property, the responsibility of a government agency. Investments to develop these assets are done with tax money, or through funds obtained or borrowed from donors. Through an open tender, the asset holder invites private sector actors to bid for lease contracts to operate the publicly owned systems under a public-private partnership construction. These private operators are expected to manage, maintain and operate the system in such a way that they recover both the costs of running the system and make some profit. The possibility they have to make a profit is assumed to provide the incentive for the operators to manage the water systems efficiently and effectively. After all, the more efficiently they operate the system, the more money they earn.

Two versions of the DMF exist in Mozambique. One hinges on FIPAG, the Water Supply Investment and Asset Fund that was founded in 1998 to develop water services in the large cities of the country. Although FIPAG is mainly responsible for major cities, it also has become responsible for some water systems in small towns. The second is the version implemented by AIAS, an organization established to mimic the responsibilities of FIPAG, but tailored specifically to support the management of water services provision in small towns. In this chapter, I elaborate how these organizational arrangements came into being, and provide details of how they make the DMF model work in practice. I compare how the two different versions of the DMF model are operationalized, each following different interpretations of the original DMF and with different outcomes.

---

[28] This chapter is partially based on: Tutusaus, M.; Cardoso, P. and Vonk, J. (2018). de(Constructing) the conditions for private sector involvement in small towns'water supply systems in Mozambique: policy implications. Water Policy. 20(S1): 36-51

1. Historical background and context of water services provision in Mozambique

The institutional framework for providing water services in Mozambique consists of a series of regulations and polices, the majority of which was espoused in the 1990s. The formulation of these regulations and policies have mainly been led by the Ministry of Public Works, Housing and Water Resources (MPOHRH). The executive arm of this Ministry is the National Water and Sanitation Directorate (DNAAS), which is the organization responsible for strategic planning and management of the Mozambican water services sector. The first water law of the Republic of Mozambique, after gaining independence from Portugal, entered into force in 1991. This law and its subsequent revisions in 1995 and 2007 were influenced by the wider political reforms that Mozambique went through during these times.

In 1987, the Economic Rehabilitation Program (PRE) (designed by the World Bank) was initiated to establish economic stabilization through the introduction of a number of policies. These policies were to allow and facilitate the development of the money market; the so-called rationalization of government subsidies and decentralization. All of these were aimed at reducing the role of the state and fostering the participation of the private sector in the Mozambican economy (IMF and World Bank, 1999). In the spirit of these reforms, the government of Mozambique also undertook 'a broad reform of water supply provision aimed at moving toward delegated management, and improving its regulation and financial planning' (IMF and World Bank, 1999). The initial 1991 framework law was later complemented and further specified with the adoption of the Water Policy of 1995. In 1995, "a great priority of the Government (of Mozambique) was to improve the state of water services, especially in urban, peri-urban and rural areas. It was necessary at that time to introduce new partners to the sector, in particular private sector operators and develop new approaches for the provision of water services"[29] (Politica de Aguas, 2007 – Resolucao n.46/2007). In line with the overall flavor of the reforms, the water law explicitly promotes the involvement of the private sector in the provision of water services and explicitly mentions private sector actors as the preferred candidates to operate water systems. Only in exceptional cases where "the private sector is not able or interested in specific settlements", does the law allow for other more flexible options, for example through the establishment of a municipal water department. In the possible absence of competent municipal authorities or a municipal council, the government can be represented

---

[29] Author's translation

in the provinces by the Provincial Offices of the Ministry (DPOHRH) and at District level by the offices of Infrastructural Development (SDPI). These last entities support the development of water services in remote areas of the country.

In 1998 the Delegated Management Framework for the provision of water services was enacted through Decree 72/98. One of the objectives of the DMF was to support and facilitate decentralization efforts in the water sector by formally allowing to grant management, lease or concession contracts to private operators. In this way, the DMF was an operationalization of the Water Policy of 1995 for urban centers. The DMF made it possible for those cities which lacked the capacity to organize their own water service provisioning to delegate this task to private operators[30]. As part of this reform, a new organization - the Water Supply Investment and Asset Fund (FIPAG) - was founded through Decree 73/98. FIPAG was to become the owner of the water infrastructure used for service provision. It had to assume responsibility for the development of new water infrastructure in urban areas, the financing of new investments and the awarding and supervision of contracts for water supply provision with private operators. Alongside FIPAG, the Water Supply Regulatory Council[31] (CRA) was established in the same year through Decree 74/98. CRA was to become responsible for the regulation of the water services and tariffs of formal water providers. It also had to safeguard the interests of the consumers.

Figure 4 Institutional Framework Water Sector Small Towns in Mozambique (Source: adjusted from WSP-AIAS, 2015)

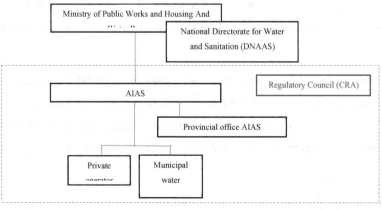

---

[30] DNA1 (June 2015) and DNA2 (November 2015)
[31] Conselho de Regulação de Água

When the DMF was established in 1998 only the five largest cities (Beira, Quelimane, Nampula, Pemba and Maputo) fell under the mandate of FIPAG. In the process of allocating these five large urban areas to FIPAG, also a number of small towns were added as they were in close vicinity of one of these larger cities. In this way, small towns like Boane and Matola became part of the service area of Maputo and the small town of Dongo was assigned to the service area of Beira[32].

As a financially-autonomous public organization, FIPAG is required by Decree to generate two-thirds of its own financial resources (for operations and investment)[33]. At the moment when FIPAG was established, the Ministry of Public Works and its executive arm, in the form of DNAAS, only considered the larger urban centers in Mozambique to have the potential of operating on a commercially viable basis[34]. This assessment was based on these cities having sufficiently large consumer market and on their favorable population densities. Hence, to allow the organization to operate as a financially autonomous organization, it was originally envisioned that FIPAG would only operate in these larger urban areas of the country. Transferring the responsibility of water provisioning in urban settlements to FIPAG was based on one or two general ideas which, although never explicitly put in writing by FIPAG, the Ministry or the World Bank, seem to have been widely accepted and agreed to by the Mozambican government and FIPAG. These were that to be considered for transfer to a private operator, a town or city needed to either have an approved investment plan or a program for the expansion of its water works, or be in the possession of a positive feasibility study showing its ability to achieve full cost recovery (at least operation and maintenance costs to begin with and later on also investment costs). These were the conditions that were deemed necessary to allow FIPAG to successfully facilitate the involvement of private operators in a commercially viable manner.

When the Government of Mozambique started to also more explicitly address water and sanitation challenges in small towns, it was reluctant to include these towns under the mandate of FIPAG[35]. This was because the poor financial viability of the systems in small towns would pose too much of a financial burden to FIPAG[36]. The concern of jeopardizing the financial

---

[32] DNASS1 (February 2016)
[33] The remaining third can be provided by the government through national budget disbursements.
[34] AIAS7 (February 2018) and DNA1 (February 2016)
[35] AIAS1 (February 2016) and AIAS6 (February 2018)
[36] DNASS1 (February 2016), DNASS2 (November 2015), DNASS3 (February 2016) and AIAS1 (November 215), AIAS2 (February 2018).

sustainability of FIPAG was shared across the board: the funding agencies interested in small towns, those agencies that had been supporting FIPAG until that date (mainly the World Bank) and the management of FIPAG itself[37]. In the words of a former high level manager of FIPAG: 'Why would you complicate a thing that is already working, FIPAG is even servicing their debts'[38]. Hence, to not compromise the (potential) financial viability of the Fund, a new solution had to be created to deal with water provisioning in small towns. In designing this solution, the government felt compelled to follow the advice and recommendations of the donors. This is logical, because water service provision in Mozambique continues to heavily rely on donor funding. Even though the financial participation of the Government of Mozambique (through the Ministry of Public Works, Housing and Water Resources) has increased up to 2016, its contribution still only accounts for a mere 20% of the total budget for investments in the water sector (UNICEF, 2017; CSO, 2010).

In 2006, the Millennium Challenge Corporation (MCC)[39] was prepared to allocate funds for the improvement of water services in small urban centers in central and northern parts of Mozambique. The MCC availed USD 5 million to "develop and apply new policies to promote sustainable management of Mozambique's water resources infrastructure"[40]. The MCC proposal included and supported the creation of a new organization, the Administration for Water and Sanitation Infrastructure (AIAS)[41]. According to those involved in the process, AIAS was established to allow for "a very quick implementation (of contracts)"[42]. The idea of quick implementation was that AIAS would not be subject to pre-existing conditions and would be allowed to set up a new system of contracts, financial arrangements or pool of operators, from scratch. In the revision of the Water Law in 2007 (Resolution 46/2007, 30 October 2007) the Government of Mozambique went along with this proposal and agreed to extend the DMF by creating AIAS. AIAS was to become responsible for the development and management of the infrastructural assets in 130 'secondary' towns (Decree 18/2009 of 13 May 2009). The arrangement was that AIAS would manage the investment plans for the rehabilitation and construction of existing and new water systems, launch public tenders to encourage the local

---

[37] AIAS1 (February 2016), AIAS5 (February 2018) and DNASS2 (November 2015)
[38] AIAS1 (November 2015)
[39] The MCC is a U.S. foreign aid agency created by the U.S. Congress in January 2004.
[40] https://www.mcc.gov/where-we-work/program/mozambique-compact (accessed 31 October 2018).
[41] https://www.mcc.gov/where-we-work/program/mozambique-compact#mz-water-and-sanitation-project (accessed 3 October 2018).
[42] AIAS4 (7 February 2018); In looking for an organization that would have exclusive dedication to such small towns, MCC seems to have ignored the role DNASS had been playing in the development of services in small towns and purposefully required the creation of a new and 'independent' organization (DNASS2 - November 2015).

private sector to participate in the operation of these systems and supervise the implementation of the contract with these operators. CRA would supervise the work developed by AIAS, control the work done by private operators and regulate the water tariff that the operators can charge. Unlike FIPAG, AIAS was not created to become a financially autonomous organization able to collect and manage its own funds because of the nature of the systems that fell under its purview.

2. The implementation of the DMF

Although the organizations of FIPAG and AIAS emerged from the same policy model and fall under the same legislation of the Delegated Management Framework, each of them has evolved differently and each of them made different types of modifications and adjustments during the process of implementation. In what follows, I describe each of these consecutively, starting with FIPAG.

a. FIPAG

Soon after its establishment, FIPAG organized an (international) tender to attract a private operator for the system of the city of Maputo, as was stipulated in the DMF. FIPAG signed the first 15-year concession contract in 1999 with an international consortium called Aguas de Mozambique for the management of the water systems in Maputo. The consortium consisted of Aguas de Portugal (Portugal), SAUR (France) and Mazi (Mozambique). This original consortium only lasted a couple of years, however. In the time of its existence, the contract underwent several renegotiations and also the partners of the original consortium were reconfigured several times (PPIAF, 2009). The reason for these changes was that the original bid on which Aguas de Mozambique had won the original contract turned out to be too low. This led to unattractive returns for the consortium, while also provoking cuts or delays in the implementation of water infrastructure works. The partners in the consortium proved to be rather inexperienced in dealing with the requirements and idiosyncrasies of implementing donor-funded investment projects (PPIAF, 2009). The new consortium without SAUR was not able to resolve the issues of underperformance in the subsequent years, consistently delivering below expectations on important targets such as coverage expansion or non-revenue water reduction (PPIAF, 2009). These disappointing results prompted the government of Mozambique (through FIPAG) in December 2010, to acquire 73% of the shares in the consortium Aguas de Mozambique from Aguas de Portugal. This acquisition of shares signified the end of the privatization era for the urban areas of the country. After acquiring 73% of the shares, FIPAG became the *de facto* operator of the urban water systems of which it also was the asset holder. This meant that the original objective of the DMF, to separate asset holding from operations, was no longer adhered to.

FIPAG used this step to further effectuate a change that had already been taken place for a couple of years. In 2008, the managing director of FIPAG realized that existing water systems were not interesting enough commercially to attract international private water operators. At the same time, the Mozambican private sector did not have the capacity to assume responsibility for the operation of water services in these larger cities[43]. This realization could have prompted FIPAG (and the Mozambican water sector at large) to have stopped the DMF, but they opted for an alternative solution. In order to maintain a structure in which operations are separate from ownership, FIPAG re-invented its own structure by establishing regional water utilities. These regional water utilities were to increasingly absorb the operational staff working for local (municipal) providers. In fact, the regional offices were an internal re-organization rather than being established as independent entities. This reorganization was possible under the DMF as the language in the regulation (Decree 73/98) stipulated that FIPAG was allowed "to accommodate operational schemes", which was read by some as providing FIPAG with the legal possibility to accommodate the actual operational part of water service provisioning[44]. The effect of this new interpretation was that FIPAG became both asset-holder and service provider[45]. As a result, FIPAG is currently involved in the operations of water services in over 15 cities (including the original five)[46], holding both the ownership of the assets as well as being responsible for operations.

In the meantime, FIPAG had been progressively taking over smaller systems that originally had not been classified as big urban centers. In the mid-1990s, the National Directorate for Water Services and Sanitation (DNASS) commissioned feasibility studies for the development of water infrastructure in 12 other major cities in Mozambique, including all the district capitals. These feasibility studies were the first step in bringing these cities under the mandate of FIPAG. In this way, in the early 2000s, the cities of Xai Xai, Chokwe, Inhambane, and Maxixe were transferred to FIPAG. The feasibility studies done for these cities advised favorably about their suitability for developing water provisioning systems in a financially sustainable manner. As the feasibility studies were being finalized, investment plans for Lichinga and Angoche were

---

[43] AIAS1 (February 2016)
[44] AIAS1 (February 2016), AIAS7 (February 2018)
[45] DNASS3 (February 2016), AIAS7 (February 2018)
[46] Interviews with AIAS1 (24 November 2015) and interview with WB2 (19 February 2016)

also completed by DNASS. As a result, these towns were also subsequently transferred to FIPAG[47]. Other towns such as Nacala, Manica or Chimoio were also transferred to FIPAG around the same period (2007-2008), even though these towns did not fulfil the implicit requirements for transfer. "It was a political decision and FIPAG rescinded some of the conditions"[48]. Although no investments had been identified for these towns, they - for example Nacala - were expected to become increasingly important in the economic development in Mozambique as part of the mining industry in the Northern regions of the country. The Nacala Corridor railway Project was to connect the port of Nacala to Malawi passing through Nampula, Lichinga, and Cuamba[49].

With the increasing number of towns covered, FIPAG has grown considerably since its inception in 1998. This size allows them to have sufficient in-house capacity in both the central office and in the regional branches, making it possible to deal with water infrastructure projects in the different towns not as individual systems operated by individual operators, but as sub-units of a larger organization. FIPAG indeed decided to organize water service provisioning in the different towns in an aggregated manner. Services that are of benefit to the entire organization are centralized in Maputo, such as legal advice, tariff analysis, procurement services and other specialized services such as project development (long-term network design). Being the operator of multiple service areas allows FIPAG to cross-subsidize investments and operations across the water systems of different cities and towns that fall under its mandate. Revenues from operations in urban centers where FIPAG has had a longer presence and that are profitable have allowed the organization to absorb systems that would initially require significant 'care' in the form of investments and re-organizations[50]. The Director of the FIPAG office in Chokwe (transferred in 2004) thus highlighted how their "system deficiencies" are still compensated by FIPAG central offices by "providing funds for investments, centrally procuring the material for network expansion works - including the material for house connections or purchasing equipment incidentally when required"[51].

Over the years, FIPAG thus has transformed into a quasi-public agency able to expand services in big and small urban centers in the country. This is not only at odds with the originally

---

[47] AIAS1 (November 2015), AIAS7 (February 2018).
[48] AIAS1 (November 2015)
[49] Accessed 29 January 2019: https://clubofmozambique.com/news/northern-development-corridor-13-years-focused-on-the-development-of-mozambique/ and https://macauhub.com.mo/2017/09/18/pt-mocambique-e-malaui-aprovam-expansao-do-corredor-de-desenvolvimento-de-nacala/
[50] AIAS1 (November 2015)
[51] FIPAG1 (February 2016)

envisioned DMF but also challenges some of the principles of commercialization that justified its establishment in the first place. By cross-subsidizing between systems it operates and because it does not feel the pressure of short-term contracts when operating these systems (as is the case for private operators), FIPAG has the freedom to take a more aggregate and longer-term view on the organization and expansion of water service provisioning. It is quite plausible that it is precisely because of this deviation from the original principles of commercialization that FIPAG is succeeding in improving service levels in the towns that it directly serves. Yet, opinions in the Mozambican water sector are divided as to the desirability of these deviations, as they contradict the original setup of the DMF. DNASS, for one, expressed their surprise at FIPAG not being invited to get involved in the expansion of services to small towns[52]. DNASS would have rather seen FIPAG absorbing the responsibilities of AIAS. CRA in turn is somewhat doubtful of the performance and financial figures that FIPAG presents[53] and blames this on the lack of transparency created by the fusion (or confusion) of responsibilities of asset holder and operator. The World Bank too has initiated steps to reform FIPAG, as they consider its current way of operating an undesirable deviation from the original model. They are particularly critical about the fact that FIPAG is both a self-managed asset fund and an operator. The World Bank has been pushing for the regulation of regional FIPAG offices where accounts can be divided. Yet, our analysis suggests that such a reform would limit FIPAG's ability to transfer resources and capacities between cities and towns and reduce the potential for cross-subsidization and limit its ability to stretch the time period for balancing revenues and costs. This would also potentially reduce its capacity to provide good service levels in the smaller towns that it serves.

b. AIAS

Since its creation, AIAS has been able to rehabilitate 18 systems[54] and has tendered 24 systems for contracts with private operators. AIAS employs 18 people directly and engages the services of 39 consultants that are financed through donor-funded technical assistance projects. Currently, it is estimated that only one-fourth of the operational costs of running the organization are covered by AIAS itself, either through internal funds or transfers from the Ministry. The rest of the AIAS budget (about 75%) come from donors[55], making AIAS

---

[52] DNASS2 (November 2015)
[53] CRA4 (February 2018)
[54] until December 2017
[55] Internal information AIAS

dependent on these donors to determine which projects they engage in and how they do so. In 2018 the main donors of AIAS were the Government of the Netherlands, UNICEF (partially funded by the Government of the Netherlands) and the World Bank. According to the first Executive Director of AIAS (2009-2010), AIAS was never sufficiently endowed to implement its mandate. The second Executive Director (2011-2018) corroborates: "the DMF would allow to progressively rehabilitate systems, but there was no financial support in terms of staff. When I came in 2011 they only had 10 staff including the person who cleans the floor and the driver"[56].

Moreover, the geographical coverage of the mandate of AIAS, with over 130 towns in total spreading to all regions of a country that extends over 7,000 km from South to North, made it difficult for staff in Maputo to effectively do their job. In order to better oversee and assist the expansion of services in the entire country, the World Bank – which through its Water and Sanitation Program had become instrumental in the institutional capacity building[57] of the organization since its foundation - started supporting the establishment of regional offices. These offices would be staffed and act as representatives of AIAS in the provinces. In 2010, AIAS established four regional offices in Cabo Delgado, Nampula, Zambezia and Inhambane to support and decongest its operations in the Head Office in Maputo. The costs of running these offices were supported by donor funds, mostly from the World Bank who initially proposed this strategy. It is the intention of AIAS that the regional offices increasingly take over the responsibilities of technical assistance and other management and administrative tasks to support local operators[58]. At the moment, however, these offices are staffed with only one or two persons and they are therefore still unable to take over responsibilities from the central office in Maputo.

Despite the vast mandate of AIAS, their budgeted expenditure for 2017 was slightly over USD 16 million[59]. Even in the event that this budget would be solely allocated to the rehabilitation of systems (which is not the case), it would only be sufficient for the rehabilitation of 4 to 5 systems per year, as AIAS estimates the full rehabilitation of one system to cost between USD 3 million and USD 4 million. The maximum budget received by AIAS was "of about USD 10

---

[56] AIAS5 (February 2018)
[57] The Water and Sanitation Program has not offered investment funds in their projects and their focus has remained to develop institutional capacity within the organization and lobbying externally
[58] AIAS2 (February 2016) and INT4 (June 2015)
[59] Exchange rate February 2018 (USD/61MZN): 1 billion Meticais (MZN)

million"[60] for 2017 for the 130 towns under its mandate. In contrast, the expenditure of FIPAG for its urban centers amounts to slightly over USD 40 million[61] (UNICEF, 2017).

The reliance of AIAS on funds from donors also means that they need to comply with the interests and project objectives of these donors. This is illustrated by their compliance with donor preferences for the way in which water systems are rehabilitated. Donors tend to have a preference for concentrating efforts on the rehabilitation of a few systems, rather than spreading funds over a larger number of towns, in which they could gradually rehabilitate the water systems. An example concerns the systems developed in the province of Inhambane by UNICEF, where the objective was to offer fully-developed systems that would be able to compete with existing (informal) private operators. Referring to this preference of donors, the former President of the Regulatory Board said that "financial institutions need to see 'nice work', put the flag of the donor in nice places, even though it is just in few places"[62]. He indicated he would prefer a diversified-investment approach as this would allow the expansion of services to more towns, even though there are no experiences to substantiate his conviction. From within AIAS, the reluctance to go for gradual rehabilitation stems from the conviction that partially rehabilitated systems are less attractive for private operators, because such systems are not (yet) reliable and profitable. Gradual rehabilitation would also turn out to be disadvantageous for AIAS as it would lead to continuous additional requests for assistance in repairs and continuous maintenance. Hence, AIAS preferred a model of infrastructure rehabilitation that would serve the purpose of creating interesting business cases to attract local private operators and allow them to implement their operations independently[63]. These systems should be able to offer sufficient scale and capacity for the operator to 'hit the ground' running and start generating revenues from the sale of water[64]. Many in the sector were critical of this strategy of AIAS, recognizing it as being primarily prompted by the need to please the donors.

The approach of fully rehabilitating water systems and then transferring them to private operators however, ran into trouble as many systems experienced start-up problems. A first type of problems related to the functioning of the water systems. In Moamba, for instance, technical difficulties with the actual flow at the intake due to the unexpected upstream abstraction of water for irrigation delayed the production of potable water for service delivery. In Bilene and

---

[60] AIAS5 (February 2018)
[61] Exchange rate February 2018 (USD/61MZN): 2.5 billion Meticais (MZN)
[62] CRA1 (25 November 2015)
[63] AIAS5 (February 2018)
[64] AIAS5 (February 2018))

Manjacaze, the operators struggled with the turbidity levels of the water sources, which were higher than assumed during the design phase. In the case of Manjacaze, AIAS was even forced to drill alternative boreholes in order to address the issue of turbidity. In the instance of Mopeia, the private operator breached the contract before the end of the five year period because of the many operational challenges and the financial implications of operating the system. In such cases, AIAS was more or less forced to remain engaged with the water systems even after they had been transferred to the private operator. Hence, AIAS remained involved in the development of the system and continued to facilitate the creation of a business case for the private operator even after the system has been transferred.

To make and keep the systems attractive enough for private operators, AIAS had little choice but to tinker with the original DMF model. They did this, among others, by implementing capacity development programs or through additional investments in the rehabilitation, upgrading or maintenance of the system. The current Head of the Technical Department of AIAS refers to this process of tinkering with the model in the following manner: "we are in constant metamorphosis. We all still have to learn what we can do"[65]. This sentiment is echoed by the former Executive Director of AIAS: "(I) do not know if what we did or decided at that time was the best. We did not know what we would find ahead. Honestly, we were experimenting. And in the process we were strongly supported by the World Bank in everything we did"[66].

AIAS is currently engaged in 24 contracts with private operators (Table 6). Each contract is linked to one single system. However, some operators have accumulated the management of more than one system. AIAS works in partnership with 16 different operators. One operator currently manages eight systems and two other operators manage two or three systems (Table 6). The other 13 are operators of one single system. The contract between AIAS and the private operators stipulates that the private operators have to pay a fee to AIAS and CRA to pay for their services and support. The private operators are supposed to pay this fee from the revenue they generate from selling water in each of the individual systems.

The first contracts that AIAS signed with private operators all adhere to a single water tariff[67] that was applicable to all systems. The reason to implement a singular tariff was that there was

---

[65] AIAS2 (June 2015)
[66] AIAS5 (February 2015)
[67] 18 MZN/m3 volumetric consumption cost or an average 23 MZN/ m3 for 10m3 consumption – domestic use- including a MZN 50 fixed charge per month

insufficient historical data to generate a tariff that would reflect the actual cost structure of each individual system. Although the first private operators with whom contracts were signed did have some experience in the water sector, none of them had hitherto operated a water system. They therefore also lacked the knowledge or reference to be able to calculate or propose a realistic tariff during the tendering process. In early 2015, AIAS performed a first evaluation of prices, based on information about operational costs provided by operators. This revealed that tariffs would have to almost double (from 18 to 30-32 MZN/m^3) in order to allow operators to (re-)cover the operational costs[68], the fee to AIAS and generate an acceptable profit. This new tariff was already discussed with CRA in 2015. However, AIAS had not yet implemented these revised tariffs in 2018 as it had not been approved by the Ministry Assembly and CRA requires additional argumentation for the establishment of a new tariff. CRA insists in establishing a tariff that does not exceed the limit established by affordability studies and the purchasing power of the users in these locations. Highlighting such limits falls under the mandate of CRA as it is to preserve the well-being of the user. Nevertheless, raising the tariffs remains one of the focus strategies of AIAS as doing so would allow them to co-fund their operations from the receipt of fees from operators, while also making interactions with operators easier. To date, none of the private operators pays the lease fees to AIAS. This is both because of the delays in tariff adjustments and because AIAS' current organizational status (legal form) does not allow it to collect payments. In the short term, new regulations concerning AIAS will allow the organization to issue invoices and manage its finances more independently[69]. AIAS was requested by decree to mobilize funds for the development of water infrastructure in the towns in which they operate, but the quantity of funds available, and the sources available have proven to hinder the capacity of AIAS to operate as effectively as expected.

---

[68] AIAS trusts the information generated by the operators, but has not been able to verify whether the reported operational costs are solely operational or if they include as well capital expenditures operators have incurred in order to expand network or other major repairs they have funded themselves.

[69] AIAS2 (February 2018); One of the main obstacles in granting this status is that such financial autonomy clashes with the interest of the Ministry of Finance to control and administer the budget of AIAS5 (February, 2018).

Table 6 Existing contracts between AIAS and private operators (December 2017)

Operator 1	Inharrime, Jangamo, Moamba, Mocuba, Mopeia (breach contract 2018), Ulongue, Morrumbene, and Homoine
Operator 2	Espungabera, Manjacaze and Nhamatanda
Operator 3	Massingir and Namentil
Operator 4	Alto Molocue
Operator 5	Bilene
Operator 6	Caia
Operator 7	Mabalane
Operator 8	Malema
Operator 9	Massinga
Operator 10	Mocimboa
Operator 11	Mueda
Operator 12	Pebane
Operator 13	Ribaue
Operator 14	Anacuabe
Operator 15	Ilha de Mocambique
Operator 16	Villanculos

In order to compensate the operators for their limited options for increasing revenues through tariff increases, AIAS has been seeking strategies to either facilitate the generation of revenue for the operator or limit their expenditure. AIAS decided to provide rehabilitated systems with a minimum number of existing household connections, to ensure a sufficiently large customer base to run operations in a financially viable manner. The initial number of water supply connections of the water systems that AIAS put up for tender ranged from 200 to 1500 (depending on funds available during rehabilitation). Hence, the idea was such a 'baseline market' would allow operators to collect revenues from the very start of their operations. After this initial start, it would be up to the operators to decide how and whether they wanted to expand to make their business more viable[70]. Expanding services would mean increasing the customer base and thus the possibilities for enhancing revenue and generating profits. In fact, because tariff increases were not possible, expanding the customer base was one of the few options for private operators to improve (financial) results. Hence, in strategizing about ways

---

[70] Viable refers in this context to the capacity of the operator to generate sufficient income so that operators would see opportunity to engage in these contracts.

to optimize their business, private operators tended to prefer increasing coverage rates (the size of the market they serve) rather than improving service levels, as more connections is a safer bet for increasing revenue generation[71]. There is a risk, however, that the expansion of services to additional customers goes at the expense of the quality of the services delivered to existing customers. This happens for instance when the expansion of the network exceeds the design capacity of the system.

As the costs for further expanding the network are to be covered by the private operator, these providers have an incentive to compromise on the quality of materials used for secondary or tertiary network expansion. Both in Caia or Bilene, operators opted for lower quality materials to ensure that the costs of extending services could be fully and quickly recovered by the new connections. The contract did not contain specific definitions of customer segments or areas that the operator was to cover. As a result, the operators could selectively expand to those neighborhoods where population densities are higher, allowing them to generate additional revenues for relatively little investment as less infrastructure is required to extend services in these more densely populated areas. This was either done at the initiative of the private operator, such as in Caia where the operator choose to expand in densely populated neighborhoods (ie. Vila, DAF, Complexo Vovo) or in coordination with the municipality, as in Bilene where the operator was requested to only extend services to those areas with 'clear roads'. There are also instances in which the operators extended services to less attractive areas at the request of the local authorities. This was for instance the case of the operator of Moamba, who extended a main pipe for over 5 km to Pessene, when another project to provide this last town with water was delayed. This is, however, not the preference of most private operators. In this case, it was directly linked to the intervention and pressure of local authorities.

AIAS has engaged in many smaller and a few more structural deviations to the original model of delegated management. Most of these deviations took place to resolve a fundamental caveat of the model: the faulty assumption that there is a solid pool of private sector operators interested in acquiring water service provisioning projects and contracts. Despite this, arguably fundamental deviation, the validity of the model has not been questioned. The approach needs to be "feasible for the operator, the Government (of Mozambique) and the people" says the

---

[71] PO1 (June 2015), PO3 (November 2015), PO5 (February 2016)

former Executive Director AIAS[72] to which the former President of CRA adds: "The government cannot reproduce fast enough the same environment that a private operator can. We are not fundamentalist about it being private operators, but it is the quickest way to develop"[73].

3. Discussion

In Mozambique, commercialization of water provisioning took the shape of the Delegated Management Framework. The two main principles that inform this Framework are (1) the separation of asset holding from operations, with the assets remaining in public hands and the operations being assumed by private sector actors, and (2) cost recovery, with the operation of water systems being expected to be financially self-sufficient in terms of generating enough income to cover costs as well as generate some profits. In Mozambique a long period of experiences with public utilities had generated sufficient doubt on the capacity of municipalities to operate these systems, but also to allocate dedicated funds to the development and maintenance of water infrastructure. The material presented in this chapter shows that even though many international policy actors consider the case of Mozambique a successful one in terms of commercializing water services provisioning (with some of them even using the Mozambique case as an example to be followed by others[74]), a closer look at how the management and operation of service provision is actually done in practice makes it doubtful whether it should even be considered a case of commercialization.

This doubt arises as during the process of implementing the model, both the asset holder and the private operators made significant changes to it. In my analysis, these changes are not implementation failures or temporary faults that will be remedied with time, but they are adjustments, additions and appropriations needed to create water provisioning systems that function in the specific context of small Mozambican towns. For the small towns that fall under the mandate of FIPAG, one important and quite significant adaptation to the model was that FIPAG gradually started combining the role of asset holder with that of operator, thereby defying one the two main key principles of the model. The reason for FIPAG to also increasingly become involved in operations was that there was insufficient interest from capable private sector actors. The first experience of the concession contract with the Aguas de

---

[72] AIAS5 (February 2018)
[73] CRA1 (November 2015)
[74] See, for example, World Bank (2017). Which presents the Mozambican case as part of its 'turn around utility' series.

Mozambique consortium showed that it was difficult to run and operate such a water system on a commercial basis. Some of the main consortium partners pulled out for this reason, leaving the Government of Mozambique and FIPAG few other options than to return operations to public hands by acquiring the majority of the shares in the consortium of operators. Combining the role of asset holder with that of operator turned out to have advantages, as it allowed FIPAG to employ a system of cross-subsidies by channeling funds from more profitable towns to support commercially less viable ones. In doing so, FIPAG also challenged another main principle of the model, that of full cost recovery. Moreover, by concentrating responsibilities and funds in one location, FIPAG was able to create advantages of scale by aggregating certain activities, such as procurement, in support of water provisioning in small town systems.

The situation for AIAS was a bit more complex as they lacked the status (and reputation) of FIPAG. Moreover, large urban cities are usually politically more important for important for governments than rural areas meaning that "public resources are disproportionally spent on [...] cities" (Bakker, 2003:333). This forced AIAS to adhere more closely to the original structure of the DMF model. However, the realities of small towns in Mozambique also prompted AIAS to make changes and adjustments. One notable change was that AIAS continued providing support to water systems even after these had been transferred to private operators. Whereas the model assumes that the contracted private operators have the capacity to operate their water system effectively and efficiently, this proved not to be the case in Mozambique. This is why AIAS decided to provide (and invest in) capacity building programs for the private sector partners that it entered into business with. Hence, AIAS engaged in training programs to strengthen the financial management and operational skills of the private operators, while also actively supporting the start-up of the water systems. The President of CRA justified this by stating: "Yes, we have the Delegated Management Framework we need to comply with, and yes, we do not have a lot of operators that can do the job but let's incentivize the private sector"! Hence, there was much more outside donor and government support for the private operator than would normally be expected under a delegated management contract. The DMF model also assumes that private operators will pay a management fee to AIAS, which would form its main source of revenue and allow it to operate in a financially autonomous manner. However, the delayed tariff increases and operational challenges have meant that the private operators are not paying such fees yet. This, in turn, means that AIAS remains completely dependent on donor and government funding.

In addition to deviating from the models, the results of implementing the model in terms of improving water access have also been disappointing. I argue that this is because of the inherent flaws in the idea of commercialization. Small towns in Mozambique are characterized by densely populated centers and sparsely populated peripheries. Moreover, because the operators do not have the right or power to adjust the tariff, expanding the market is the only option they have to increase revenues. Private operators, who are also responsible for the expansion of services, tend to focus on the more populated areas, where they are best and most quickly able to recover costs. Apart from this spatial concentration of network expansion, such extension in network extension is also characterized by a temporal dimension. As the duration of the delegated management contract is limited, private operators have an incentive to frontload investments in expansion. The result of this is that private operators expand services early on in the contract, prioritizing the increase of coverage over improvements in service provided to existing customers. In addition, private providers also have an incentive to use mediocre or sub-standard materials as this lowers investment costs and allows these costs to be recovered more quickly. The assumption is that these materials will hold for the duration of the contract. The effect of this is that there still is a substantial portion of the population in the peripheral areas of the small towns who have no access to the water provided by the private operator. Moreover, the use of sub-standard materials raises questions about the sustainability of the water systems in the near future.

Interestingly, relevant stakeholders, such as CRA, World Bank, FIPAG, and AIAS argue that the reason why expectations of the DMF model in terms of improving service levels and increasing access have not been met in small towns is a result of the faulty or partial implementation of the model. As this is a diagnosis that leaves the belief in the model intact, it is one that is favored by many with stakes in water service provisioning in Mozambique. The overall reluctance to question the model, and the principles of commercialization on which it is based, clearly shows in the very creation of AIAS. At the time AIAS was created, FIPAG was already performing a dual role (as the private sector actors lacked capacity and interest) and the operational difficulties of the Maputo concession were already apparent. Hence, even though the original DMF was not adhered to and even though results were disappointing, donors and government agencies insisted in pursuing an elaboration of the DMF for the governance of water provisioning in small towns by establishing AIAS.

This adherence to the model can partly be explained by the almost dogmatic belief in commercialization, and in the idea that the operation of water systems is most effectively and efficiently done by the private sector. When the local private sector was not able to provide the operators envisioned by the model, the reaction of AIAS and donor agencies was to invest in and develop these operators through a full-fledged program of capacity development. Ironically, it was the lack of capacity in the public sector (municipalities) that was the argument to expand the DMF to small towns in the first place. However, the response of developing the capacity of municipalities to deliver water services, was clearly not an option for the actors involved[75]. This points to a pre-existing discontent among donors and others involved with the functioning of the public sector in Mozambique, while also re-confirming the conclusion of chapter 2 of this thesis that commercialization was and is an undisputable paradigm in international policy and donor circles. Hence, that operation of water systems is best done by the private sector is a given, and something that Mozambique needed to buy into if they wanted to continue to receive donor support. Another possible explanation for the choice to expand and replicate the DMF model through AIAS is to view it as the result of institutional path-dependency: establishing AIAS was relatively easy as the original blueprint and overall legislative framework for the organization were already in place. Not only did the regulations documenting the DMF exist, but also donor and government agencies were acquainted with the model: they had become accustomed to speak, think about and plan the provisioning of water services following the language and ideas of the model. The only thing that had to be done was extend the framework beyond the urban areas and ensure that the extension of the model would not negatively impact FIPAG. This meant creating a special agency for the small towns for which the DMF had been extended.

For AIAS, as the asset holder, the model justifies the existence of the organization. Without a DMF that stipulates the separation of asset holding and operations, there would be no reason for AIAS to exist. In fact, the delegated management model has become a fundamental part of the organizational identity of AIAS and of those working for it (Tutusaus et al., 2018). Moreover, the financing of AIAS is strongly dependent on donor agencies that have associated themselves with and strongly believe in the success of the DMF. It is therefore in the interest of AIAS to portray itself as a dedicated follower of the DMF, as this will likely ensure continued access to donor funding. Donors likewise depend on the success of DMF for their own reputation and legitimacy. For donors such as the World Bank and the MCC, the pursuit of

---

[75] AIAS6 (February 2018);

commercial principles in water provisioning is an important cornerstone of their WASH policy. The DMF, which is has been presented as a successful case of delegated management (World Bank, 2009), not only demonstrates their pursuit of these principles but also allows these donors to showcase the success of their approach. This is perhaps best illustrated by a recent World Bank publication on the DMF framework in Mozambique, which is part of the "utility turn around series" (World Bank, 2017: 1). For the Government of Mozambique, it is in their interest to promote the donor-supported model stipulated in the Water Law (and subsequent amendments) as it allows them to both claim progress in achieving increased access to water services (with the ultimate aim of universal service coverage), whilst at the same time relieving themselves of the financial responsibilities of expanding coverage.

The persistence and continued promotion of the DMF along-side the adjustments to the model as it is implemented by both FIPAG and AIAS, highlights a discrepancy between the model-on-paper and its implementation in practice. The model and its implementation exist in two different political domains. The DMF exists in the global and national policy domain, where it is promoted by donors and Mozambican water sector agencies. The implementation is what comes into being through the everyday actions and adjustments of FIPAG, AIAS and the private operators. These are geared towards making water systems work within the overall political, financial and institutional context of Mozambique. The reputation and continued existence of organizations like FIPAG and AIAS importantly depends on their ability to suggest enough overlap and convergence between the two models to convince both the donors as well as the government as well as the customers. They do this by re-defining DMF – FIPAG inventing regional water utilities is a case in point – or by using reporting formats that make it seem as if they diligently follow the model. In Mozambique this is more challenging for AIAS than for FIPAG, as for AIAS the divergence between everyday operational realities and idealized model is bigger. The irony is that while AIAS remains dependent on donors and the government for a large part of its funding, it needs to conceal this dependence in order to remain a legitimate organization in the eyes of these same donors. At the same time, AIAS needs to make all kinds of concessions to the private operators it works with to keep them interested. Reconciling and juggling these different domains and different forms of legitimacy requires a lot of creativity.

Chapter 6: Commercial Public Utility in Small Towns in Uganda[76]

In this chapter I trace the history of the National Water and Sewerage Corporation of Uganda (NWSC) and document how it functions in small towns. I show how NWSC successfully responded to donor pressures to privatize or commercialize. Today, NWSC sees itself and is seen by others as a successful example of commercialization. Yet, when zooming in on the details of the organizational and financial arrangements for water provisioning in the Ugandan small towns that NWSC is responsible for, it becomes clear that NWSC has had to adapt some definitions and principles of commercialization to make these systems work. The NWSC needs to creatively navigate political and donor pressures, whilst trying to improve the number of people benefiting from NWSC water. This is done by re-inventing commercialization to make it into something that can be combined with cross-subsidization and a dependency on external support. The chapter is another illustration of how attention to the detailed operational details of water provisioning is important when trying to understand what works and what doesn't.

1. Historical background and context of water services in Uganda

Following a recommendation of the African Development Bank (AfdB), NWSC was established in 1972 (Decree No. 34) as a parastatal or corporatized utility. As a corporatized utility, NWSC would remain a government-owned organization while enjoying autonomous corporate status under a special law or act in which tasks, responsibilities and powers invested in the utility would be reflected (Braadbaart et al., 1999:9). The idea behind this was that the autonomous status of this organization would allow NWSC to operate outside of the influence of politicians while remaining in the public sector. Yet, as Braadbaart concludes on the basis of a review of this model, 'most corporatized utilities were autonomous on paper, but not in practice' (Braadbaart at al., 1999: 10). This was also reported to be the case in Uganda where NWSC was run "analogous to political-family businesses. [...]. Service expansion was politically driven" (Mbuvi and Schwartz, 2013: 379). This may explain why also the benefits expected from corporatization, such as improvement of the utility's financial situation with

---

[76] This chapter is developed based on data collected by Maxi Julius Omuut for the completion of his MSc thesis research phase (*Original thesis title: 'The Impact of Infrastructure Development on Water Supply Services and Viability of Small towns'*) at IHE Delft in 2018. I was his mentor in this process. I collected additional data in Uganda myself. Parts of this chapter have been used to develop an article: Tutusaus, M.; Schwartz, K. and Omuut, M. (submitted). Commercialization as Organized Hypocrisy: The Divergence of Talk and Actions in Water Services in Small Towns in Uganda. Water Alternatives.

increased billings, improved collection rates, higher staff productivity[77] and ultimately an expansion of the quantity and the quality of services, did not materialize. As Schiffler (2015:144) reports: "in Kampala there was an imbalance between greatly enhanced water treatment capacity and water connections, which lagged behind. The Corporation billed only half the water it produced, and of the amount billed, it collected only 60%. It had far too many staff for a company of its size. Staff costs accounted for 64% of the total operating cost. Its debt was too high and […] senior management did not really empower mid-level managers".

In the second half of the 1990s, the debts of NWSC amounted to US$ 53 million, leaving the organization on the verge of bankruptcy (Muhairwe, 2009). The situation in which NWSC encountered itself became one of the justifications of the reforms rolled out by the World Bank more widely in the country towards the end of the 1990s[78]. The timing of these reforms in Uganda coincided with the privatization decade and the promotion of private sector participation in public services[79]. Between 1998 and 2001, the German engineering company, H.P. Gauff Engineers, was contracted to run the Kampala Revenue Improvement Programme (KRIP) for a period of three years (NWSC, 2003). This contract was signed under pressure from the World Bank and in accordance with the reform program. This private sector participation contract with Gauff Engineers was seen as a first step in a "transition to full-fledged privatization" (Muhairwe, 2009:12). At the same time, Dr. William Muhairwe took office as the new Managing Director of the NWSC. His "aim quickly became to show that NWSC could improve its performance without private sector participation, or in other words, that there was a viable public sector alternative to privatization" (Schiffler, 2015:147). In resisting privatization, the top management of the NWSC sought support from a broad coalition of actors both internally (Board of Directors, Area Managers, staff, etc.) and externally (media, Uganda Public Employee Union, senior officials in the Ministries) to the organization (Mbuvi and Schwartz, 2013). Muhairwe also tried to downplay achievements of the management contract by claiming that it was "largely a failure" (Schiffler, 2015). As an alternative to full privatization, Muhairwe proposed the commercialization of the water provider. This alternative consisted of the introduction of management principles and practices such as a strong efficiency-orientation, competition, performance management and entrepreneurialism.

---

[77] Staff productivity is usually reflected as a ratio of the number of staff per 1,000 connections. Tynan and Kingdom (2002) suggest that this ratio should be lower than 5 staff per 1,000 connections.
[78] For more details about these reforms see Schiffler (2015) and Mbuvi and Schwartz (2013)
[79] In 1997, NWSC was listed for full privatization by the Ministry of Finance and the Local Government Act Cap 243 and the Water Act Cap 152 were revised and enacted to allow private sector participation in the provision of water services (Mbuvi and Schwartz, 2013).

These are principles and practices that are usually associated with the private sector (Mbuvi and Schwartz 2013: 380)[80]. Implementing these changes would entail a thorough revision of the corporate culture and the mindset of the staff working at NWSC, which according to Muhairwe (2009) posed the greatest obstacle to turning around performance in the utility (Schiffler, 2015; see also Buyi and Schouten, 2010).

The introduction of the rapid change management programme in the utility improved the performance of NWSC significantly between 1998 and 2004 as highlighted in table 7. NWSC expanded services by over 25% in three years, increased billing collection to 98% in the same period, and reduced non-revenue water from 49% to 33.3%. These performance improvements strengthened the bargaining position of the utility vis-a-vis external (donor) agencies, as well as the responsible Ministries (Finance and Water and Environment) in Uganda (Mbuvi and Schwartz, 2013). While these changes were taking place in Uganda, the donor community, and especially the World Bank, was already looking for an alternative to the traditional privatization approaches they had until recently tried to push through. The hopes of improving operational efficiency, increasing capital investment and increasing access rates during the privatization decade had failed to materialize and proved to be unrealistic. Moreover, the attempts of governments to introduce private sector participation in the provision of water sector was increasingly becoming a controversial political issue, and in some cases led to unrest (Prasad, 2006). The alternative proposed in Uganda by Muhairwe addressed the concerns and ideas of the large donors in the water sector (i.e. World Bank) as it worked with and emphasized the benefits of private sector practices, yet maintained water in the public domain.

Table 7 NWSC Performance 1998-2010 (Source: Mbuvi and Schwartz, 2013)

Indicators	1998	2004	2010
Unaccounted-for-water (as a percentage of total	49 %	37.6 %	33.3%
Staff productivity (total staff per 1000	36	10	6
Service coverage	48 %	65 %	74%
Connections (total active water supply accounts)	34,272	100,475	246,459
Collection efficiency	71 %	98 %	98%

The successful transformation (Table 7) of NWSC in this period made 'commercialization' an important part of the identity of the NWSC within the country, in the region and worldwide.

---

[80] Although the commercialized public utility has a long history (see Blokland et al., 1999), the link with the turn-around in performance in "one of the poorest countries in the world" (Schiffler, 2015:159) made NWSC the 'textbook' example of the commercialized public utility.

Members of the NWSC management travelled the globe to present their success story at international conferences such as the Water Week at the World Bank and the Corporation was repetitively "showered with awards" (Schiffler, 2015: 154). During their presentations, staff of the National Water and Sewerage Corporation emphatically presented their organization as "a successful example of a commercial public utility that combines public sector control with private sector efficiency" (Muhairwe 2009). Its Five Year Strategic Direction identifies financial growth and sustainability as one of the four strategic priority areas[81] (NWSC, 2016). Similarly, the corporate profile defines the utility as a corporation that is structured and run to "to operate on a commercial and viable basis" (NWSC, 2018). The annual report 2016/2017 highlights how "the Corporation registered [...] an unprecedented operating profit before depreciation of Uganda Shs 70 billion[82]" (NWSC, 2018:23). The same annual report also triumphs the Corporation's application of sound business practices. Till today, NWSC maintains this self-representation as a successful commercialized public utility. Combined with a continued emphasis on their capacity to implement what they call 'turnaround programs', this has proven to be an effective strategy to continue receiving support from governmental agencies in Uganda, as well as from international donors for funding[83].

a. The expansion of NWSC's mandate to small towns

When the NWSC was established in 1972, the utility was responsible for the supply of water to the three main cities of Kampala, Jinja and Entebbe. By 2010, this mandate had been expanded to 23 urban centres[84], which amounted to approximately 58% of the Ugandan population (Kitonsa and Schwartz, 2010). Through this expansion in the mandate of NWSC, the utility became responsible for the cities and large towns (settlements with more than 15.000 inhabitants) of Uganda. Small towns, falling in the category between 5,000 and 15,000 people[85], were managed directly by Municipalities who were free to engage with private operators

---

[81] The strategic priority are of financial growth and sustainability is then operationalized through 8 'key deliverables': enhancing viability of new towns, value for money investments, investment financing, income diversification, cost optimization and efficiency, revenue growth, compliance and governance and integrated ICT solutions (NWSC, 2016).

[82] About US$ 18.87 million (exchange USH/USD)

[83] Schiffler (2015) argues that there are some doubts about the actual performance of NWSC. Instances of debt relief by the Ugandan government in 2008 and queries about the accuracy of some of the performance figures produced by the NWSC have raised questions about how successful the turn-around of NWSC really is.

[84] See also Kitonsa and Schwartz (2010) for a discussion on the transfer of towns to NWSC. Similar to the ambiguity of allocating towns in Mozambique, also in Uganda the transfer of towns to NWSC seems to be rather ambiguous.

85 In 2008, 14 local private operators managed water systems in 62 small towns in Uganda (Kitonsa and Schwartz, 2010).

through management contracts under the supervision of the Ministry of Water and Environment.

Figure 5 Institutional Framework for Water Services in Uganda (Source: Kitonsa and Schwartz, 2010)

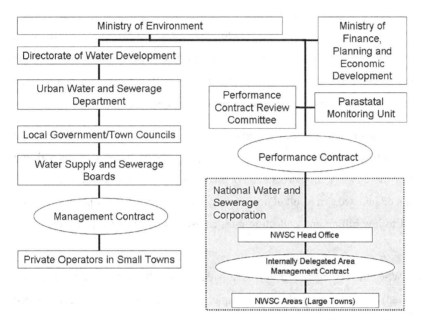

In 2013, the Ministry of Water and Environment of Uganda decided to enlarge the mandate of NWSC to progressively also absorb the management and operations of small towns, villages and other settlements in transition such as rural-growth centers. The arrangement for water services in these settlements was to transition progressively from municipalities engaging in individual contracts with local private operators to branch organizations operating under the umbrella of the NWSC. The reason for this was that the arrangement between municipalities and private operators had led to discontent at the Ministry. Systems were not being maintained or were abandoned, funds were not allocated transparently to the benefit of improving operations and access rates did not improve[86]. As a result, by February 2018 the mandate of NWSC was expanded to include over 260 additional towns. In what follows, I elaborate on the process of how NWSC has incorporated these towns in its organizational structure and what impacts this has had on the utility as a whole, as well as on the individual systems.

---

[86] Hq9 (May 2017), NG1 (May 2017) and LG1 (May 2017)

4. *Organizational structure and strategic directions of NWSC for small towns*

In the terminology of the NWSC, once it officially takes over the management of a specific system, that system becomes a 'Branch'. NWSC devised a protocol to cluster several systems (or Branches) in the vicinity of each other into one Area. Most of the operational work that requires daily contact with the final users and the local authorities, such as billing and revenue collection, takes place at the Branch level. The Area level office is staffed by a manager who is responsible for overseeing billing and collection activities of the Branches and an engineer who supports the technical work of the different Branches in that particular Area.

The NWSC is aware of the challenges of managing and expanding services in small towns. The most important challenges when they took over water provisioning in these towns had to do with the state of the existing infrastructure. For many small towns, the infrastructure was reported as being dilapidated as a result of poor maintenance by the previous operator. This was partly attributed to low billing and collection rates as users were not used to paying for the service[87].

According to NWSC, the aggregation of towns in Areas would allow for the service provider to develop (limited) economies of scale. For example, "instead of having an engineer for each town, we have one engineer managing over 5 towns in a cluster form. Therefore the labor costs of this engineer are shared across several towns. Management proficiency is what is sustaining the operations of these towns"[88]. One such clustering initiative is in Bushenyi town, which is located in the eastern region of Uganda. NWSC took over the management and operations of the Bushenyi Town Council water supply system in 2002. Since then, the Town Council was elevated to municipal status in 2010 with the annexation of the Ishaka Town Council[89] to form the Bushenyi/Ishaka municipality (Bushenyi-Ishaka.Municipality, 2015). This annexation also led to an expansion of the service area with Bushenyi/Ishaka operating under the NWSC. At a later stage, NWCS added 10 towns (Kabwohe, Bugongi, Rubirizi, Kitagata, Kabira, Kashenshero, Mitooma, Rutookye, Kyabugimbi and Buhweju to the Bushenyi/Ishaka operational area to form what is now known as the 'Greater Bushenyi/Ishaka Cluster'.

Decisions taken at NWSC Head Office significantly shape the functioning of the utilities in the operational areas, with the Head Office devising policies and regulations to guide the operations

---

[87] Hq9 (May 2017)
[88] Hq1 (November 2017)
[89] Part of the annexation included a few adjacent parishes.

in the areas. The Head Office also controls the financial accounts for the revenues collected from the Areas and remits monthly management fees to help cover the operations and maintenance costs of the water supply systems in the Areas. They also provide technical support and funding for the development of infrastructure. Yet, the Areas are given some degree of autonomy to make decisions regarding service delivery and asset development. For example, they themselves identify the need for investing in the extension of mains; they carry out feasibility studies and also budget for the mains extensions. However, for implementation they need approval from the Head Office. Indeed, reviews show that the Head Office has, in fact, rejected, downsized or postponed some of the infrastructure projects proposed by managers of Areas (and Branches) in the past.

The expansion of the mandate of NWSC to small towns was accompanied by a clear policy direction of the Government of Uganda. The National Development Plan 2015-2020 (NDPII) and the strategic Vision 2040 of the ruling government demanded an increase in safe water coverage from 71% (in 2015) to 90% by 2030. NWSC incorporated this ambition in their Five Year Strategic Directions documents of 2013-2018 and 2016-2021. The flagship program of the Corporation is the 100% Service Coverage Accelerated Project (SCAP100), which envisions to reach 100% service coverage in urban areas by 2020[90]. This explicit ambition implies that considerable efforts need to be dedicated to the expansion of infrastructure in small towns.

The Head Office regulates the operations in the Areas through internal performance contracts referred to as Performance Autonomy Creativity Enhancement[91] (PACE) contracts. Performance targets and standards are annually negotiated and agreed upon between the Head Office (regional representation) and the Areas. The targets stipulated in the PACE contracts translate to the obligations NWSC is committed to through the 3-year performance contract[92] the Corporation holds with the Government of Uganda (Mugisha et al., 2007). This performance

---

[90] SCAP100 aims to meet the national goal of the Government of Uganda for the water supply and sanitation sector of increasing urban and rural access of water supply services. The current coverage levels are 77% (urban) and 65% (rural) and these would be increased to 100% and 79% (minimum of one clean and safe water source per village/town) by the year 2020 (NWSC-SCAP100, 2016).

[91] The PACE contracts were previously referred to as Internally Delegated Area Management Contracts. See Mugisha et al. (2007) for a discussion on the usefulness of such contracts for performance improvements in the water utility.

[92] The 3-year performance contracts between the NWSC and the Government of Uganda were introduced by the NWSC Act, which came into effect in 2000. The 3-year performance contract largely mimics management contracts which are established with private operators under a public-private partnership arrangement (Schiffer, 2015). The fifth performance contract (so-called PC5) ran from 2015-2018 (NWSC, 2018)

contract is revised every three years and updated to reflect the most pressing challenges[93]. The requirements established at central level are subsequently cascaded down the organization until they reach the Branches. Operational Areas are directly held accountable for all daily operations such as the establishment of a comprehensive asset management system to 'improve business continuity', undertaking 'effective programs' of service reliability improvement, and implementing cost reduction campaigns such as non-revenue water (NRW) reduction measures (Omuut, 2018). These PACE contracts simulate the contracts the Head Office already held with Areas within the main cities (i.e. Kampala and Jinja had been divided into Areas with its respective Area Managers). The managers at Branch and Area level are held accountable for the performance of their systems and rewarded accordingly, either through the provision of a fee for management or with the disbursement of funds to support the development of the 'business'[94].

5. Adjustments to the idea of commercialization

The performance of Branches and Areas is roughly measured by and equated to their financial performance. NWSC expresses the performance of water service provisioning in small towns in what is called the working ratio, which represents the relationship between (operating) costs and revenues. A working ratio above 1 "means that the utility fails to recover even its operating costs from annual revenue" (Tynan and Kingdom, 2002: 3). A ratio below 1.0 indicates the utility is recovering its operational costs and some of its capital costs. The World Bank uses a figure of 0.68 as the target for the working ratio. For a water utility to be considered financially sustainable its working ratio needs to be 0.68 or lower (Tynan and Kingdom, 2002). NWSC uses a different target, as it is mainly interested in the capacity of each individual Branch system to (re-)cover the operating costs from the revenues generated by that very same system. Here NWSC thus deviates from what the World Bank and other international policy actors prescribe and promote[95]. Unlike them, NWSC does not expect its Branch or Area offices to recover non-operational costs, such as the depreciation of the assets or interest payments of debts incurred to develop those assets. NWSC can afford this deviation because it operates on the basis of the expectation (based on past practice) that infrastructure development will be either subsidized by the GOU directly (as stipulated in the 3-year performance contracts), or be funded through

---

[93] Hq6, Hq7 (November 2017)

[94] Workshop Kampala February 2018.

[95] As highlighted earlier, Franceys et al. (2016:79) suggest that the costs that need to be recovered are "investment costs (both hardware and software, with the implication of a subsidized 'cost of capital') to install a WASH service (CapEx + CoC); regular, recurrent operating and minor maintenance costs (OpEx) plus major maintenance and renewal costs (CapManEx – again both hardware and software)".

development partners[96] (GOU-NWSC-PC5, 2015). Since 2013, the GOU has indeed allocated Ushs 3 billion[97] each financial year for the expansion of infrastructure in small towns to be used at the discretion of NWSC. In addition, the GOU contributes financially to other programs, such as SCAP100[98]. Project funding by development partners happens on a more *ad hoc* basis. NWSC also allocates internally generated funds for the development of infrastructure through the so-called 'infrastructure fund' (NWSC, 2015)[99]. Through this infrastructure fund the development of 'deficient' systems is cross-subsidized (NWSC, 2015).

Depending on the source of funding used, NWSC has been able to maintain an average overall working ratio of 0.78 (NWSC, 2018) for the 2016/2017 financial year, thus enabling the utility to recover their operational costs. However, when disaggregating these figures it becomes clear that "most towns prove not to be viable according to this definition of financial sustainability"[100] (NWSC-SD, 2016). NWSC in fact uses the surpluses generated by typically larger 'well-functioning' systems such as Jinja, Mbarara or Entebbe to cross-subsidize those systems that are not able to generate sufficient revenues to cover their own O&M expenses. This is perhaps best illustrated by looking at regional variations in the Areas of NWSC. The Kampala region, which covers the metropolitan area of Kampala and incorporates close to half the connections operated by NWSC, has a working ratio of 0.54 (NWSC-CP, 2018), whereas Bushenyi only had a working ratio of 1.16 in 2016-17 (it was even lower in 2013-14). Only through cross-subsidization from the more profitable large urban areas is the NWSC able to maintain an average working ration of below 1 for the country as a whole (Table 8).

Table 8 Working ratio for NWSC (Omuut, 2018)

Financial years	FY 12/13	FY 13/14	FY 14/15	FY 15/16	FY 16/17	Av. WR
NWSC Global	0.80	0.83	0.96	0.84	0.78	0.84.2

As mentioned above, the NWSC still mainly relies on the traditional means of financing infrastructure which include revenues from water sales, GOU subsidies and donor-funded investments. As part of its attempt to become more commercial, the NWSC has however also expressed its desire to access investment funds through commercial loans from the local capital market (NWSC-SD, 2016). The ability to access other forms of funding besides donor funding

---

[96] Hq6,Hq7 (November 2017), Hq8 and Hq9 (December 2017)
[97] About US$ 815,000
[98] Hq1-Hq9 (November and December 2017)
[99] Hq4 (November 2017)
[100] Hq4 (November 2017)

is becoming increasingly important for a number of reasons. First of all, it is important as the NWSC acknowledges that accepting donor funds also implies aligning with the agenda of the donor. Moreover, donor funding is somewhat unreliable and often *ad hoc*. Accessing other funding sources thus reduces the dependency on donor financing. A second reason for wanting to access other sources of funding is that the ability to do this would further contribute to the image of the Corporation as a successful and viable commercialized public utility. Indeed, being able to access market finance would provide evidence that the water utility is 'bankable'. Illustrative here is a recent article in the Global Intelligence Water Magazine entitled "Creating Africa's Most Bankable Utility", in which the current managing director of NWSC explains how the utility is seeking "to tap into new ways to fund its investment drive" (Global Water Intelligence, 2019). The concept of 'bankable utilities' is also becoming increasingly important for donors, who see bankability as an indicator of the longer-term sustainability of their own funding (Baietti et al. 2006). Because of this, and somewhat paradoxically, convincingly portraying the NWSC as a 'bankable utility' may also be part of a strategy to please donors and thus continue accessing their funds, rather than necessarily implying the actual intention of obtaining investment funding from the market. Indeed, showing the ability and possibility to access market finance is itself becoming an indicator of being a commercially viable utility. This explains why the management of the NWSC gives such prominence to its financial status in the annual reports. The annual report of 2010, for example, highlights a "[r]ating of A2 and a long-term rating of A, which when translated implies that the Company's liquidity factors and Company fundamentals are sound, and that the risk factors in regard to commercial borrowing are low" (NWSC, 2010).

Tellingly, the emphasis on its favorable credit rating and the emphatic expression of the desire and ability to access commercial loans has not yet translated into the actual diversification of sources of investment funding. As the NWSC has been established as a commercial public utility, it is to operate on the basis of cost recovery. In theory, it should also be able to access and pay for its own funding. As shown, however, the practice so far has been that NWSC borrowed money from the GOU and experienced difficulties to repay this. In 2006, the managing director of the NWSC, Dr. Muhairwe, justified this by arguing that full cost recovery in developing countries was "a myth"[101] (Schwartz, 2008). He stated that full cost recovery would require tariffs that ordinary Ugandan citizens could not possibly afford (Schwartz, 2008; Schiffler, 2015). Based on this argument, Muhairwe requested the Government of Uganda to forgive the loans (valued at US$ 90 million) that it had on-lent to the NWSC (Schiffler, 2015). Rather paradoxically, this was done by arguing that this would make the NWSC more "commercially-oriented" as the debt-reduction would make it easier for the utility to take out loans from the local capital market. The NWSC did also initiate plans to take out US$ 18 million in loans through the bond market[102]. Because of the global financial crisis, however, "the Ugandan Ministry of Finance stopped the bond issue from going ahead, citing the need to first use conventional concessional financing sources. Although foreign loans were denominated in hard currency, their overall conditions – longer maturity and lower interests – were considered better than local bond financing. But these loans would be borrowed by the government and – as opposed to the previous practice – would not be on-lent to NWSC" (Schiffler, 2015: 156). Hence, in spite of how it tries to portray itself to the outside world, market finance has not played a role in the actual financing of NWSC investments (Schiffler, 2015).

a. The details of financial performance in the Branches

The Bushenyi/Ishaka cluster had a working ratio of 1.07 at the end of the financial year 2016/2017, meaning that the cluster was still not commercially viable in that year. For the town of Kitgum the working ratio was 1.16 in that same financial year[103]. These figures are no exceptions, but resemble those of the majority of small towns operated under the auspices of NWSC (Omuut, 2018). The biggest cost for the utilities are employee-related costs, next to maintenance of the treatment plants and the piped network. For Bushenyi, these costs accounted for 48% (employee costs) and 23-28% (maintenance) of the total operating expenses in the FY2016/2017 (Table 9).

One of the main difficulties in achieving financial sustainability for small towns lies in the way that NWSC sets its tariffs vis-a-vis the operational costs. The Head Office of NWSC applies a uniform tariff to all its operational areas. The argument they use is for doing this is that such a uniform tariff promotes the equitable provision of water supply services across the country. This uniform tariff is designed to allow operators to recover O&M costs plus depreciation and some surplus for investments at a national scale. The operational areas of NWSC are then tasked, given this uniform tariff, to meet the O&M costs as a minimum requirement through their water sales revenue[104]. The implication of this construction is that the operational areas

---

[102] The bond issue was assisted by the World Bank, which supported the NWSC in its endeavour to access funds through the bond market (Schiffler, 2015).

[103] This means that for every US$ 1 received in revenue, the utility has US$ 1.07 or US$ 1.16 in operating costs.

[104] Hq1,Hq4 (November 2017) BI1, BI3 (December 2017)

are not able to independently set tariffs for water and sanitation services on the basis of the real local costs of operations.

Table 9 Expenditures and Billing 2014-2017 for Bushneyi/Ishaka and Kitgum (Source: Omuut, 2018)

	Bushenyi/Ishaka Cluster				Kitgum Area			
	FY 13/14	FY 14/15	FY 15/16	FY 16/17	FY 13/14	FY 14/15	FY 15/16	FY 16/17
Employee related costs	579,234	932,792	1,315,712	1,359,712	421,976	492,435	474,878	482,802
Premises and maintenance	26,510	57,360	75,794	103,769	32,937	19,093	21,223	29,097
Static Plant and Pipe maintenance	366,528	530,929	727,019	650,850	155,829	188,713	230,525	277,898
Transport and Mobile plant	30,687	80,052	93,458	103,300	28,101	49,116	45,746	52,348
Supplies and services	126,618	219,872	319,686	316,895	25,027	17,749	16,710	35,442
Administrative expenses	102,837	215,499	222,894	282,691	99,936	118,256	125,479	129,648
Total Operating Expenses	1,232,414	2,036,504	2,754,563	2,817,365	763.806	885,362	914,561	1,007,235
Billing	1,097,276	1,257,464	2,789,408	2,644,450	307,651	544,517	764,595	870,787
Working ratio	1.12	1.62	0.99	1.07	2.48	1.63	1.20	1.16
Unit cost of production	3,371	3,939	3,236	2,770	5,429	3,678	3,530	3,722
Water tariff[105]	2,422	2,263	2,668	2,855	2,422	2,263	2,668	2,855

Studies have confirmed that operational costs for water provisioning in small towns surpass those in bigger urban settlements (World Bank, 2017). Due to low economies of scale, economies of density and operational inefficiencies, small towns have higher costs of production per unit of water in comparison to large and medium urban centres (World Bank, 2017). The collected evidence from Uganda seems to confirm this generalization. The performance trends from the financial year 2013/2014 until 20515/2016 show that the unit cost of producing a cubic meter of water in the Bushenyi/Ishaka cluster and Kitgum varied between

---

[105] NWSC uses a uniform tariff for all systems. Tariffs are revised (generally annually, sometimes bi-annually) to reflect inflation rates, domestic and foreign price index and electricity tariff.

3000 Ushs/m^3 and 4000 Ushs/m^3. During the same period, the average cost of production used by the NWSC Head Office fluctuated between 2500-2900 Ushs/m^3. Illustrative in this context is the comparison with the unit costs of Kampala. For the financial year 2015/2016 the unit cost of production in Kampala was 1,961 Ushs/m^3, which is approximately 60% of the unit cost of production in Bushenyi/Ishaka (3,236 Ushs/m^3) and 55% of the unit cost of production in Kitgum (3,530 Ushs/m^3). In Kampala, the existing demand allows for the full exploitation of the existing water infrastructure, which makes it possible for utilities to achieve the most efficient production rates as per design. At the same time the ratio of connections/km is higher in Kampala (Table 11). Bushenyi/Ishaka cluster and Kitgum operate their treatment facilities below full capacity. Particularly in the case of Kitgum, the treatment facilities appear to have been designed with a vision of expanding demand for water services that has yet to materialize (Omuut, 2018).

Table 10 Treatment plant utilization (Source: Omuut, 2018)

	Bushenyi/Ishaka Cluster				Kitgum			
	FY 13/14	FY 14/15	FY 15/16	FY 16/17	FY 13/14	FY 14/15	FY 15/16	FY 16/17
Plant capacity (m^3/day)	2,000	3,104	3,233	4,521	2,539	2,539	2,376	2,376
Production (m^3/day)	1,476	1,952	3,244	3,317	472	866	921	841
Utilization	74%	63%	100%	73%	19%	34%	39%	35%

The figures presented show that the transformation of these systems into financially sustainable ventures is challenging. Although Management of NWSC is aware of this, NWSC as a corporation maintains the position that small towns can be commercially viable undertakings for the Corporation (recovering O&M costs) in the long run. According to managers at the Head Office[106] positive results can be achieved if the 'right' investments are made: "from NWSC experience, we have taken over towns before that were argued not to be financially sustainable… for example, Arua, Bushenyi[107] and Soroti. By the time we took over they were very small towns, Bushenyi had about 500 accounts and billing about Ushs 5,000,000, Arua

---

[106] Hq3 (November 2017)
[107] The reference to Bushenyi here is to Bushenyi town, before it had been clustered into Busheny-Ishaka Area

had 1,000 accounts and billing Ushs 3,000,000. But thanks to the right investments – those that focus on the expansion of networks and the increase of the customer base - Arua, Bushenyi and Soroti are now billing about Ushs 230,000,000[108]".

b. Strategic development of infrastructure in small towns

A significant proportion of the operational costs in small towns is related to the type of infrastructure used to expand services in these towns. The choices in water production facilities and the quality, length and location of the distribution network are of strategic importance in determining cost efficiencies. The national standard design manual of Uganda distinguishes between different types of towns. For instance, standard designs for rural areas target a per capita consumption of 20 l/day for a town with a population of up to 5,000 persons. Medium towns with a population of up to 20,000 inhabitants have design standards for a per capita consumption of 35 l/day and larger towns with more than 20,000 inhabitants should adhere to designs allowing for a per capita consumption of 50 l/day (M.W.E-Manual, 2013: 2-4). This distinction in demands per type of town would also justify a distinction in different types of infrastructural choices for each category. Yet, NWSC does not distinguish between towns in terms of infrastructure and implements the same standards for infrastructure development everywhere. Usually the choice consists of a combination of a piped network for the central areas in town, which are generally more densely populated, and standpipes in more remote areas of the town. These choices are guided by costs, with the cheapest possible arrangement usually being preferred. When external funders are involved in the development of projects, other criteria may outweigh costs, such as long term savings against immediate gains (i.e. installing bigger pumping mains to augment water supply in the small satellite towns to avoid drilling boreholes in the future)[109].

In the approach that NWSC has adopted for taking over small towns, the development of water infrastructure development takes on a particular prominence. Since 2014 it implements two plans, the Infrastructure Service Delivery Plan (ISDP) and the Water Supply Stabilization Plan (WSSP), which both focus on the rehabilitation and construction of water supply systems. The plans target activities such as the upgrading of treatment plants, the rehabilitation of transmission mains and the piped distribution network, the expansion and extension of the

---

[108] This is an estimate of average monthly billing (per town). The example also highlights that even within a cluster such as the Bushenyi-Ishaka cluster there may be (smaller) pockets where the provider can recover its operational costs.
[109] Hq6 (November 2017)

sewerage network, and the construction of public stand pipes. By thus emphasizing infrastructural development, NWSC sets out to both meet the target of ensuring universal service coverage, while also addressing the poor working ratios of these small towns (NWSC_SD, 2016). However, the development of infrastructure in strategic locations in these towns has a direct impact on the costs of providing services and the capacity of the Branch office to generate revenues. For example, in each, newly-acquired small town, NWSC chose to invest in expanding the piped network by at least 5 kilometers. NWSC did this to, first of all, meet the commitment of universal service coverage that the Government of Uganda has made in its National Development Plan 2015-2020 (NDPII). The NWSC is currently the main operating vehicle for the Government of Uganda to achieve this target. NWSC also prioritizes network expansions because it aspires to improve the commercial viability of the Corporation, which it considers as critical for its longer-term sustainability. The importance of commercial viability and the ability to recover costs means that NWSC prioritizes infrastructural development in those places where costs can be recovered relatively easily or quickly[110]. Hence, the potential customer base, their economic capacity and their consumption patterns largely determine where the 5 km of infrastructure is developed and what kind of system is put in place.

Despite their efforts to reconcile the commercial and social obligations of the Corporation the two objectives of service and coverage expansion and commercial viability are not always compatible. In fact, the importance for both the GOU and NWSC to show progress in access and coverage has obliged the utility to initially compromise on their commercial considerations. According to the Infrastructure Service Delivery Plan (ISDP) NWSC developed 1,448 kilometer of piped network in the financial year (FY) 2014/2015 (Table 11). This is a major change from the average annual network extension of 80-90 km that NWSC had developed in the years prior to the introduction of this plan.

Table 11 Network extension and connection expansion 2013-2017 (Omuut, 2018)

KPI	FY 12/13	FY 13/14	FY 14/15	FY 15/16	FY 16/17
Network expansion (km/yr)	86	470	1448	888	911
New connections per yr	21,637	28,068	34,165	38,836	43,214
Total connection	317,292	366,330	415,838	472,193	529,402
Ratio new connection/new km pipe	251	59	24	43	47

---

[110] Hq9 (May 2017), WK1 (February 2018)

Table 11 also shows the additional challenge NWSC is facing in expanding services in small towns. Due to the relatively low population density in small towns, the ratio of connections/km decreases. Whereas in the financial year 2012/2013 only 86 kilometer of additional network were required for 21,637 new connections (252 new connections per kilometer of new network), this figure dropped sharply with the gradual transfer of small towns to NWSC from 2013 onwards. In the past three years the number of new connections per kilometer of network expansion decreased from 252 (financial year 2012/2013) to about 50 (financial year 2016/2017). The decrease was particularly strong in the financial year 2014/2015 when only 24 new connections per new km of pipe were constructed.

These less favorable conditions for NWSC, however, do not mean that NWSC has been curtailing their efforts to expand services. The investments in infrastructure did translate into an increased service coverage from 77.5% to 78.5% (NWSC, 2014). The strategy for the expansion of networks coincides with other strategic developments. First, the rapid expansion of infrastructure, and associated expansion of services, signals their compliance and willingness to implement the mandate imposed by the Ministry of Water and Environment of pursuing universal service coverage. Despite the burden these efforts place on the Corporation's performance, it also provides the NWSC with crucial political backing to access investment financing from the Ministry and the Government of Uganda (Mugisha and Berg, 2017) for present and future water projects. The earlier highlighted dependency on government subsidies to finance capital investments require the NWSC to heed the requests of the government Ministries. Second, NWSC also uses the visible manifestation of its presence in many towns through its investments in rapid infrastructural growth to flag that they are committed to deliver better services[111]. Staff of NWSC argue that being positively visible in this way makes it possible for NWSC to introduce a water tariff and require payment for services, something that many users were not acquainted with[112]. According to NWSC, previous service providers "were not as diligent in collecting revenues"[113] but they were also not diligent in providing reliable services. The reputation of being a reliable service provider, even if it is still not possible to reach all consumers, lays the basis to move towards a system that encourages payment, which in turn allows NWSC to move to a system that can recover the costs of operations.

---

[111] LG1 (May 2017), Hq5 (January 2018)
[112] Hq9 (May 2017), WK (February 2018)
[113] FGD1 (February 2017)

The emphasis on the financial performance of the Branches encourages the water operators to prioritize the expansion of services to those locations which are relatively cheap to reach, so that revenues will be relatively quick and high[114] (see also WSP, 2009). Increasing service coverage, of course, also allows to increase billing incomes because of the larger number of connections that can be charged for water consumption. At the same time, extending services to the most densely populated areas in the towns allows for a higher ratio of connections per kilometer of network, which improves relative returns. These two considerations – increasing billing income and improving income per kilometer of network - largely determine how NWSC pursues financial viability. The effect is that where the first years of the cluster are characterized by the rapid growth of the water infrastructure system (and associated investments in infrastructure), the strategy NWSC adopts once it is present in a town is geared towards ensuring financial sustainability of the existing system in the area. Currently, investments to expand the hours of service are limited as priority is given to network expansion, despite NWSC's claim that it has the intention to increase their service hours to 12hrs/day, and ultimately provide services 24/7[115]. For example, in the Bushenyi/Ishaka cluster the actual investments greatly exceeded those initially budgeted for in 2014-2015 (Table 12). However, the annual customer satisfaction surveys conducted from 2015 to 2017 for that cluster reveal that customers were least satisfied with water supply reliability and the pressure at which water was delivered (Omuut, 2018). This is an indication that the system is becoming over-stretched and that the investments were most likely concentrated in the expansion of the system (through kilometers of pipe and connections) rather than in system improvements.

Table 12 Budget realization Bushenyi/Ishaka (Omuut, 2018)

	FY13/14	FY14/15	FY15/16	FY16/17
Budget (Ushs million)	44,377	129,445	317,500	185,640
Actual implemented (Ushs million)	68,460	1,829,764	1,179,709	689,497

From the financial year 2015/2016 onwards, capital expenditures on the piped network were reduced. Both the Kitgum and the Bushenyi/Ishaka cluster illustrate these trends by the lower number of network extensions and connections (which peaked between 2013/2014 and 2015/2016). These developments highlight the strategy of NWSC of first investing funds into

---

[114] See also Castro and Morel (2008:192) who argue that "utilities in Africa tend to target 'high priority' areas for expansion where immediate financial returns are more promising" (see also Schwartz et al., 2017').

[115] Hq1,Hq2 (November 2017) BI1 (December 2018)

the development of the piped network at the start of transfer to then focus on the returns on investments already made in order to make the Area/Branch more financially sustainable.

6. Discussion

In this chapter I have examined how NWSC in Uganda implements the model of the commercialized public utility. The chapter shows that there is a striking difference between how NWSC portrays itself as an exemplary case of a commercialized public utility to the outside world, and the image that appears when zooming in on the everyday practices of small-town operators who form part of the NWCS model and approach. Below, I present the details of this difference, and discuss what it means both in terms of the effective and efficient organization of water provisioning in small towns and in terms of the viability of the model of the commercialized public utility.

a. Questions about commercial viability

NWSC´s expansion of services in small towns in Uganda has proven to be more challenging than the expansion in other, larger, urban areas. Despite these challenges, the Head Office of NWSC insists that it is possible for small towns to be commercially viable, provided adequate investments in water infrastructure are made[116]. In order to convince others of this opinion, the NWSC likes to point out the improvements and progress towards financial sustainability achieved in towns such as the Bushenyi/Ishaka cluster. Although the improvements in Bushenyi/Ishaka appear promising, and perhaps realizable in the coming years, it is questionable if the NWSC can replicate these results in other small towns. Each and every small town under NWSC is different, and similar interventions would obviously not deliver the same results. The specific characteristics of each town influence the degree to which it can achieve financial viability and the current figures seem to indicate that the expansion is posing challenges to NWSC. The comparison of the Bushenyi/Ishaka cluster with that of Kitgum is a case in point. These two towns present two very different business cases. Over the past four years, Bushenyi/Ishaka and Kitgum received different investments for network expansion. In Bushenyi/Ishaka an additional 368 km of network was developed, whilst in Kitgum an additional 32km was constructed. Moreover, the total water production per consumer is approximately four times higher in Bushenyi/Ishaka than in Kitgum where 28 liters per

---

[116] Hq3 (November 2017)

consumer are produced. This suggests that Bushenyi/Ishaka serves a much bigger market than Kitgum. At the same time, the cost of production in Bushenyi/Ishaka is 75% of that in Kitgum. The specific characteristics of the Bushenyi/Ishaka cluster (market characteristics, costs of production) are therefore much more favorable for achieving commercial viability than the characteristics in Kitgum.

The difficulties of realizing commercially viable water systems in small towns in Uganda have been documented before. In a study by Price and Franceys (2003) on the financial sustainability of water systems managed by private operators in five towns in Uganda, the authors reveal that only two out of the five towns were able to function in a "financially sustainable manner". The main obstacle to realize commercial viability, according to these authors, was the low water sales, "as a result of household incomes and high water tariffs" (Price and Franceys, 2003: 21). Therefore, NWSC's reference to Bushenyi/Ishaka to show the commercial viability of water service provision in Ugandan small towns can, perhaps, best be interpreted as a way to promote a particular model by linking this model to one particularly successful example[117]. Bushenyi/Ishaka offers the ingredients and conditions required to be made into a showcase example of how to expand services while 'preserving' commercial sustainability.

There is another aspect of the service modality in Ugandan small towns that raises questions about the commercial viability of such systems. A feature that characterizes both the Bushenyi/Ishaka cluster and Kitgum is the overdesign of the infrastructure for providing services. Particularly the treatment facilities have a capacity that is not fully utilized. This problem of infrastructure overdesign is not unique to small towns in Uganda. As also noted by Franceys et al. (2016:84) "There is a steady trend to over-invest in small towns ahead of that town's ability to service any investment. […] This makes it difficult for any small town utility to operate in a conventional manner and suggests that there needs to be support to recurrent maintenance costs as well as capital investment, a charge which is unlikely to be recoverable through tariffs". Indeed, overdesigned water systems leading to the underutilization of production capacities have been noted elsewhere (WaterAid/BPD, 2010; Mugabi and Njiru, 2006; Adank and Tuffuor, 2013; Moriarty et al., 2002). The 'future demand' that systems are designed for is rarely realized in these towns. Such errors in projections may turn out to be costly to the utility in terms of investment costs, when the forecasted connections do not materialize (Hopkins et al., 2003; Lauria, 2003).

---

[117] See Rap (2006: 1312) for the need of organizations to perform 'successes by showcasing successful pilot cases.

b. The need to aggregate and cross-subsidize

NWSC cushions the impact of the challenges discussed above through its organizational structure. Because the small towns operate under the flag of a national utility, some of the (financial) challenges facing the water systems at Branch level can be accommodated. Firstly, the national reach of the water utility allows it to aggregate individual service delivery systems. The clustering of various water systems in the Bushenyi/Ishaka cluster is an example of that. Whereas some operational activities need to be kept at the individual system level, others can be centralized to the cluster or even the regional level, allowing for economies of scale. Without a national mandate, such a shifting of tasks between levels would be much more complicated. Secondly, NWSC makes use of its 'national' mandate to cross-subsidize between operational areas and water systems. The NWSC as a whole presents a working ratio of 0.84, suggesting that the utility is able to recover operational costs and can contribute to part of the capital costs for infrastructure development. In achieving this working ratio, the large urban centers play a crucial role. Given the uniform tariff that NWSC employs, the financial gains made in large urban centers allow for the provision of cross-subsidies to other, smaller systems. Most illustrative here is the city of Kampala. With its working ratio of 0.55, Kampala *de facto* functions as a 'cash cow' for the Corporation, allowing it to support the investment needs of those areas that are not able to recover costs[118].

The possibility of realizing economies of scale by aggregating towns and the practice of cross-subsidization make it possible for NWSC to present water systems in small towns as commercially viable. Yet, it is important to clearly distinguish at what level this commercial viability is achieved and the mechanisms through which it is achieved. The operational details presented in the chapter shed some serious doubts about the possibility to achieve full cost recovery and commercial viability for all small towns. NWSC may be able to demonstrate that at a national level it can operate as a commercially viable public utility (in terms of covering operation and maintenance costs), the chapter shows that at the level of individual small towns or small town clusters some level of subsidy and external support will remain necessary for many. In this analysis it, thus, becomes important to distinguish the scale at which a particular model is being analyzed.

After a period of infrastructure investment emphasizing development of the piped network in the town, attention is shifted towards guaranteeing reliability of water supply in the town. This

---

[118] Hq4 (November 2017)

is important as it helps the organization to shift back to their concern of achieving, or striving for, commercial viability. In this second phase the Branch Office becomes increasingly wary about expenditures and turns to low cost options for expanding services. During this phase capital expenditure on the piped network is gradually reduced and the focus of infrastructure works shifts towards improvements that are geared towards realizing more efficient operations, such as rehabilitation of the treatment plants to enhance efficiency in production or network upgrading to improve reliability of services and reduction of losses. This choice of investments puts more emphasis on service reliability of existing (and paying) customers over increased access, potentially limiting the access to water services to those who are not yet served (see also Mugabi and Njiru, 2006).

c. Commercial viability and the dependence on subsidies and grants

The experiences with the expansion of services in small towns in Uganda presented in the chapter suggest these will only be effective and successful when significant modification to the original models of commercial viability and autonomy are allowed for. NWSC knows this, as it is itself actively involved in making some of these modifications. One example of these modifications is the slightly adjusted definition of commercial viability (or financial sustainability) used by NWSC. To the operator a small town is sustainable the moment it is able to recover the costs of operations from the revenues generated in that same system. Ideally, Branches are encouraged to cover for minor investments. This interpretation of financial sustainability, however, is quite different from the idea of full cost recovery, where capital expenditure and operational expenditure are recovered.

Yet, in its communications to the outside world NWSC never abandons its focus on and belief in the idea of commercial viability, or in the model of the commercialized public utility. In the case of NWSC, the emphasis placed on commercial viability is a reflection of the dependency of NWSC on external actors. In particular, the roles of the Government of Uganda and external donors are important in this respect. The utility needs to navigate the national political landscape in Uganda to ensure that it retains (financial and political) support and it needs to ensure that it retains access to funding from bilateral and multi-lateral donors. In this sense, NWSC needs to operate in multiple (political) domains simultaneously. It needs to engage with the donors and the national government in the global and national policy domains, whilst it needs to provide actual services in the everyday domain. These domains are not completely separate. A modified version of the commercialization model provides guidance to how the NWSC operates in the small towns.

In this landscape of inter-dependencies between national and international actors, the initial emphasis that NWSC places on infrastructure development in small towns is crucial. This approach, apart from targeting commercial viability in the long-run, provides immediate political support for the organization. By showing service expansion in (semi) rural areas and contributing to the policy lines of the Ministry of Water and Environment, the NWSC maintains national political support for the organization. This political support is important, because if, as Dr. Muhariwe has argued, cost recovery in developing countries is a 'myth', then the NWSC will remain dependent on the financial support of the Ugandan government to cover its costs for a considerable time to come.

At the same time, by presenting itself as a successful commercial public utility, NWSC ensures its continued access to resources from multilateral and bilateral donors. Meyer and Rowan (1977:352) highlight that "organizations which exist in highly elaborated institutional environments and succeed in becoming isomorphic with these environments gain the legitimacy and resources needed to survive" (Meyer and Rowan 1977: 352). In creating the appearance of being isomorphic with its institutional environment, NWSC's active performance as a successful commercial public utility plays an important role. By demonstrating its adherence to important elements of this model, such as cost-recovery, autonomy and performance management, NWSC increases its popularity and legitimacy in the international donor community and provides the showcase Ministry, donors and NWSC require. Indeed, in its external portrayal, the NWSC manifests itself as a commercial public utility, one that targets cost-recovery, performance management and customer-orientation, diligently displaying those features that the international water policy community considers to be characteristic of well-performing water utilities (see for example Baeitti et al., 2006). Without such a portrayal, accessing donor funding would be much more complicated for the NWSC.

d. The embodiment of the commercial public utility

Through the national, regional and international portrayal of NWSC as a successful commercially-oriented public water utility, the identity of the utility has become almost synonymous with the model. In explaining the success of irrigation management transfer, Rap (2006:1313) highlights how organizations "co-produced a vast stream of promotional materials that conveyed the success of the policy model…" A similar phenomenon can be witnessed in the case of NWSC. A large number and wide variety of awards have been bestowed upon NWSC over the past years (Schiffler, 2015). These awards, which are displayed prominently on the NWSC website[119] vary from the best African water utility (2016), to Uganda's most compliant procurement entity (FY 2013/2014), to the African Water Leaders Award (FY 2014/2015). Linked to the international recognition through awards, the utility is also prominently present in international water organizations and at conferences. The current Managing Director of NWSC (Dr. Mugisha) is a Board member of the International Water Association. The Director of Business and Scientific Services, Dr. Rose Kaggwa, is on the IWA Women in Water Committee and vice-president of the African Water Association Scientific and Technical committee. A stream of supporting academic and practitioner-oriented publications have been produced over the past 15 years. Noteworthy in this stream of literature is the role played by two Managing Directors (Dr. William Muhairwe and Dr. Silver Mugisha). The title of Muhairwe's (2009), *Making Public Enterprises Work: From Despair to Promise - A Turn Around Account,* largely reflects the main topic of the book. The most prolific contributor is, however, Dr. Mugisha[120], whose articles -- such as 'State-owned enterprises: NWSC's turnaround in Uganda' (2008), 'Using internal incentive contracts to improve water utility performance: The case of Uganda's NWSC' (2007) -- highlight the technocratic practices of successfully 'turning-around' the performance of NWSC in accordance with commercial principles.

NWSC has become, so to speak, the champion and expert of the model of commercialization. The organization has become the very embodiment of the commercial public utility model in developing countries and in this way it has actively participated in the further promotion of this model. Despite the challenges currently faced with the expansion of the mandate to small towns, they continue to actively embody the model. It is noteworthy that the promotion of NWSC as a 'champion' is as much of interest and use to the utility as it is to the international community of donors and financiers. For international donors it is important to be able to support a 'successful' water utility that – apparently – adheres to a model of water service provision that prominently incorporates the dimensions (cost-recovery, autonomy) that they promote. This model justifies their donor programs and the 'successful portrayal' of it allows these donors to continue promoting the model to a broader audience of utilities and governments. To do so, these donors speak about the model in rather generic terms. For these actors it is less important

---

[119] https://www.nwsc.co.ug/index.php/about-us/accolades
[120] SCOPUS (www.scopus.com) counts 18 publications by Dr. Mugisha. Accessed 23 May, 2018.

that the model needs to be re-interpreted on the ground, or more importantly, that it clashes with the very same principles that underpin the original model

Chapter 7: Community Based Organizations in Small Towns in Indonesia [121]

In this chapter, I describe and analyze a third form that commercialization can take: community based organizations or CBOs. This model was adopted for the organization of water provisioning in small towns and rural areas in Indonesia. Within the boundaries of Lamongan Regency, on which case this study is developed, approximately 250 CBOs operate water supply schemes. These water systems service approximately 85,000 household connections, which provide water to an estimated 500,000 people in Lamongan Regency. Between 70%-80% of the total Regency's population (Planning Agency, 2016) are served by CBOs. Despite the efforts of central, regional and local governments to develop water services through local government-owned public enterprises (PDAMs), the formal utility (PDAM) only serves 15,000 household connections (approximately 75,000 people). I use the chapter to present the case of CBOs in the Lamongan Regency, focusing on the strategies these organizations have developed to operate in a commercially viable manner. Like with the examples of commercialization presented in the previous chapters, CBOs in Indonesia had to creatively adjust some of the principles underlying the CBO model for it to work in the context of Lamongan Regency communities.

1. Historical background and context of water services in Indonesia

The institutional framework governing water services provision in Indonesia is stipulated in Law 7/2004 on Water Resources and Government Regulation 16/2005 on Water Supply System Development. According to these regulations, the authority for water services development and delivery rests with the local government, which has to establish government-owned companies known as *Perusahaan Daerah Air Minum* (PDAMs). These PDAMs have the status of a parastatal. They are expected to operate as commercialized public utilities, functioning at arm's length of the government: they should be politically autonomous and financially independent by operating on the basis of full-cost recovery (Hadipuro, 2010; ADB, 2012). In actual practice,

---

[121] This chapter is developed based on data collected by Riski Aditya Surya for the completion of his MSc thesis research phase (*Original thesis title: 'Commercialization of Community Based Water Service Providers: the case of Lamongan Regency Water Service Provision*) at IHE Delft in 2017. Mireia Tutusaus was his mentor in this process. Parts of this chapter have been used for a forthcoming publication: Tutusaus, M.; Schwartz, K. and Surya, R. (forthcoming). Degrees and forms of commercialization: community-managed water operators in Lamongan Regency, Indonesia. Water.

most PDAMs face considerable challenges, many of which originated during the 1997 Asian financial crisis[122] which made debt restructuring slow. The Asian Development Bank estimated in 2012 that 79% of the 319 Indonesian PDAMs were "heavily indebted" (ADB, 2012:9). Moreover, many PDAMs have received little financial support from local governments to help them make the necessary investments in infrastructure[123] (Hadipuro 2010; World Bank 2012; AUSAID, 2013). To this day, PDAMs remain dependent on funds from provincial governments and particularly on funds from the Ministry of Public Works and Public Housing to develop water infrastructure. Further compromising the principles of commercialization, PDAMs tend to be strongly influenced by local government politics: the local government has a strong influence on the appointment of the managing director and on the setting of tariffs, often keeping these artificially low (ADB, 2012:9). The political appointments and low tariffs both negatively influence the operational performance of the utility.

The Government of Indonesia aspires to achieve universal water service coverage by 2019[124] (ADB, 2012). Given their low performance, it is estimated that the PDAMs will only be able to service 40% of the population[125]. The remaining 60% of the population will thus need be serviced through other water providers. The legal framework does provide the possibility for the local government to engage in a cooperation contract with legal entities other than PDAMs, as long as it is outside the service areas of PDAMs. There are three possible service provision arrangements in small towns in Indonesia: services through a PDAM, services through a CBO or services provided through a partnership between the PDAM and the CBO (BAPPENAS, 2003). The Government of Indonesia has nevertheless made it clear that it views PDAMs as the preferred option for water services delivery. As a result, this provision modality receives more political and financial support than the others. This is visible in the government budget which is mainly geared towards supporting the PDAMs (Surya, 2017).

---

[122] In 1997, many PDAMs had debts financed in foreign currency (mainly US$). With the drastic devaluation of the Indonesian Rupiah during the financial crisis these debts became unserviceable (Hadipuro, 2010; World Bank, 2012)

[123] The unwillingness of the local government to invest in water services is particularly visible in the sanitation sector. Throughout Indonesia there are only 11 municipal sewerage schemes, which have all been constructed by the central government (AUSAID, 2013).

[124] In 2015, it was estimated that reaching universal service coverage for water supply by 2019 would require an annual investment of US$ 4.7 billion. This required amount is 4.5 times higher than the actual investment in water supply infrastructure (WSP, 2015).

[125] The ADB (2012) notes that service coverage by PDAMs has declined in Indonesia from 39% in 2000 to 31% in 2010. The ADB suggests that the decentralization policies initiated in 1999 are an important factor in explaining this decline in service coverage.

e. The CBO model in Indonesia

Indonesia has delegated the development of water supply systems in low-income and rural areas to communities. This largely follows the international policy trend on community managed water supply systems and is also supported and funded by external aid agencies such as the World Bank and Australian Aid (AusAid). In order to allow the Ministry of Health and Ministry of Public Works[126] to accommodate these donor funds, the government of Indonesia established the 'National Policy on Community Based Water Supply and Sanitation Services'[127] (National Planning Agency, 2003). The formal possibility of delegating water supply services to community-based organizations has led to a surge in community-based water provisioning in Indonesia. Estimates of the number of CBOs that operate in the Indonesian water services sector vary from around 13,000 CBOs (WSP, 2015) to "tens of thousands" (WSP, 2011:2). The World Bank alone has developed 9,000 rural supply schemes in Indonesia between 1990 and 2010 (WSP, 2011b).

An important factor that explains the popularity of CBOs for water provision in Indonesia is that the prospect of commercially-viable, community-based enterprises is appealing in view of their potential to reduce the financial burden on financially-strapped local governments (WSP, 2011). According to the policy on CBOs, these organizations are to be governed according to the following principles (BAPPENAS, 2003):

— Water services need to be seen and treated as an economic good. The policy highlights the need to change "existing public perceptions" that water "has no economic value". Increasing "awareness and understanding of water's status of an economic good, overall practices of water use will improve, leading to a reduction in exploitation, and increased efficiency in use and a greater 'willingness to sacrifice' to obtain water". In making reference to water as an economic good, the policy explicitly refers to 'Dublin-Rio principles'.

— CBOs should adhere to a demand-responsive approach. The community should assume "the role of decision-maker in the selection, financing, and management of the [water supply and environmental sanitation] system".

---

[126] The Ministry of Public Works would later be renamed Ministry of Public Works and Public Housing.
[127] The Policy notes assistance of the World Bank's Water and Sanitation Program (WSP-EAP) and the Australian Government's AusAID (BAPPENAS, 2003).

— The Government should act as a Facilitator. This implies that the government's role, particularly at the municipal and district level, is to provide technical and non-technical advisory services, with an aim to empower the community.

— CBOs should operate on the basis of cost-recovery. Because the government is reluctant to make more funds available for water supply, it anticipates that future development and management of water supply and environmental sanitation infrastructure should be "based on the principle of cost recovery, meaning that all financial components of development (such as budget planning, physical construction, O&M, and depreciation) have to be considered and accounted for". The principle of cost-recovery is forwarded as crucial in order to "ensure sustainability".

Although the Indonesian government is heavily dependent on CBOs for achieving its aim of universal service coverage by 2019, first experiments with the CBO model were disappointing. A WSP (2011) study, which examined five districts in West and East Java, shows that one in four CBOs established in this region were no longer operational at completion of the project. A subsequent report by the WSP (2015) highlights that community-based organizations face many challenges. These include (WSP, 2015:14):

— The lack of capacity to maintain and expand services;

— A lack of access to finance, as commercial banks remain reluctant to provide loans; and

— An unclear legal framework in relation to the expansion of services across more than one village.

f. The introduction of commercial CBOs in Lamongan Regency

The initial approach of the World Bank to extend services through CBOs in the 1980s focused on infrastructural development and the technical capacities of service providers through the project "Water and Sanitation for Low Income Communities" (WSSLIC)[128]. In the 1990s, a second generation of World Bank-sponsored CBO projects (PAMSIMAS) was implemented in Indonesia which focused more on 'institutions and incentive structures'[129]. In Lamongan Regency, these projects in collaboration with the Indonesian Infrastructure Initiative (INDII -

---

[128] WB1 (February 2017)
[129] The Water and Sanitation for Low Income Community (WSLIC) project - Word Bank – and later on the follow-up 'National Rural Water Supply and Sanitation' (PAMSIMAS) project resulted in the establishment of 'thousands' of small water systems operated and maintained by CBOs. The program targeted eight provinces covering 2,461 villages from 2000 to 2009.

AUSAID), launched a program to develop CBOs (locally referred to as HIPPAMs) resulting in more than 250 small scale systems in the Regency.

In order to help the HIPPAMs achieve 'long-term sustainability', the projects provided training to HIPPAM staff on financial management, assistance with the development of technical and financial proposals, in addition to training on the operation and maintenance of water systems. In line with the policy principles, CBOs were encouraged to strive for financial sustainability by being rewarded with a performance grant (repayment of half of the loans) upon meeting agreed performance indicators (WSP, 2011). The role of the local government was to create 'enabling conditions'. The establishment of a particular HIPPAM, its functions as well as the composition of its management committee, were to be stipulated in a 'village regulation'.

HIPPAMs were expected to function as completely autonomous organizations, which would be accountable to three main stakeholders: the village chief or village government, the local government and the consumers. To account for their actions to the village and local governments, HIPPAMs were expected to present monthly financial statements and an annual report to both the local government and the village chief. As for the consumers, HIPPAMs had to hold a meeting at the end of each year inviting the consumers to discuss the report presented by the HIPPAM committee. While the HIPPAMs thus report to the community, they are not formally accountable to the final users. Users who participate in the forum can only provide advice to the management of the HIPPAM.

The organization structure of a HIPPAM is generally quite simple. A typical HIPPAM consists of a committee of paid staff of 6-10 persons. The operation and maintenance of the water system are done by volunteers from the community. HIPPAMs are organized under an association that acts as the umbrella for all HIPPAMs in an area. In Lamongan Regency, the establishment of the association of HIPPAMs emerged in the post-construction era of the WSLIC project (World Bank) in 2004. In the beginning, a WSLIC Program Consultant guided the association. It only became an independent organization in 2009. Initially, the main objective of this association was to foster the sustainability of the HIPPAMs (built by the WSLIC project) by facilitating coordination and exchange of experiences. It also conducted consultancy assignments on the management of drinking water and sanitation for the HIPPAMs (LP3ES, 2007). This role shifted as the association became self-funded, and thus needed to secure revenues: the association became more 'business-oriented' and started seeking to provide services and advice to the HIPPAMs in exchange for a service fee. Hence, the current HIPPAM Association provides technical assistance and supports the procurement of materials, but does not make any direct financial contribution to individual HIPPAMs. Although the project envisioned that local governments would play a facilitating role helping individual HIPPAMs become successful, in practice the local government did not offer any political, technical or financial support to the HIPPAMs. The local government also explicitly articulated their wish to see HIPPAMs operating independently. As HIPPAMs could not request or depend on public support for their development, they had to define their business plan accordingly.

The projects under which the HIPPAMs were set up provided the initial investment that covered the construction of the basic systems. The systems introduced by the international donors (World Bank and AUSAid) were low cost, consisting of relatively basic technologies: a well with a pump, a tower reservoir, network pipelines and metered household or yard connections. Most systems do not apply treatment or filtration of the water. Apart from the initial investment, network extensions and system improvements (i.e. additional pumps or reservoirs) were to be developed fully by the HIPPAMs themselves, with only sporadic assistance from the HIPPAM's Association. HIPPAMs in the Lamongan Regency generally access groundwater resources, with only two examples which use surface water resources.

Figure 6 Water production and distribution scheme - HIPPAMs (Source: Surya, 2017)

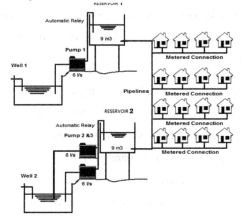

Each HIPPAM can establish its own water tariff depending on what costs they need to recover (i.e. servicing debts, account for asset depreciation, operations and maintenance) and the ability of their users to pay. The tariffs are presented to the Village Chief and the users. Although neither the Village Chief nor the users have the formal power to approve or veto the proposed tariffs, the management of the HIPPAMs sets the tariffs after due consultation with the community, making sure there is consensus and understanding. It is remarkable that the

HIPPAMs have achieved relatively high billing and collection rates (80%-98%) and relatively high profit margins. Depending on the HIPPAM, profits ranged from 15% up to 57% (Table 13). HIPPAMS have used these profits to expand service coverage in the water supply landscape of Lamongan Regency. In some cases, HIPPAM are requested to pay a contribution fee to the community for the exploitation of the water resource. Generally this is expressed as a percentage of the total revenue (usually about 3%).

2. The operationalization of the CBO model in Lamongan Regency

Mainly because they received little external financial support, HIPPAMs in Lamongan have been forced by the donors and the government to strive for commercial sustainability. On the face of it, they have succeeded. Yet, the survival and good performance of the HIPPAMs in Lamongan cannot in itself be taken as evidence that the HIPPAMS work according (or because of adhering) to the principles of commercialization promoted by both donors and the Indonesian government. A closer look reveals that HIPPAMS creatively adjusted these principles in order to create an organization that better suits the cultural and community values pre-existing their establishment.

a. Securing revenues

The Lamongan Regency provides a favorable bio-physical and institutional context for HIPPAMs to develop and function. For one, the Lamongan Regency is located in an area where easy access to clean water is limited, as the water from local surface water bodies and shallow groundwater wells is considered to be of poor quality. Moreover, the public water utility is not reaching the majority of the population and certainly not the more rural areas. This has prompted the appearance of private (informal) water vendors, which provide water in parallel to the services offered by the HIPPAMs. However, the private water vendors charge a much higher price for the water they sell. They offer water at IDR 2,000[130]/jerry can (20 liter) or IDR 100,000/m^3. The average water tariff of the HIPPAMs ranged from IDR 750/m^3 to IDR 2,000/m^3, or 50 to 133 times lower than the prices charged by informal vendors. The CBO tariff is even lower than the PDAM tariff of IDR 2,700/m^3. Hence, the HIPPAMs are succeeding in providing water services at a tariff that is attractive to consumers. The lack of alternatives also means that HIPPAMs are generally welcomed with open arms.

---

[130] €1 = IDR 16,888 (26 July, 2018); HPM7A (November 2016)

Table 13 Overview characteristics eight HIPPAMS cases (adjusted from Surya, 2017)

Data Overview	Unit	CBO							
		Tlanak (2007)	Kemlagi Gede	Geger	Karangwedoro	Sukomulyo	Trepan	Pawer Siwa	Pengumbula-nadi
Tariff	IDR/m3	1300	1000	750	1000	2000	1300	2000	2000
Billing collection rate	(%)	95	95	80	95	80	90	85	98
Average Revenue	IDR million/ month	32.43	11-12	12-13	12	12-13	9-10	18	5-7
Average Cost	IDR million/ month	20.85	6	7	7	10-11	7-8	14.5	2-3
Monthly Gross Profit	IDR million	11.55	6	6	5	2	2	3.5	4
Operating Ratio [131]		0.65	0.50-0.55	± 0.60	0.60	0.85	0.75 – 0.80	0.80	0.40
Numbers of committees	Persons	6	9	10	9	9	8	8	6
Average Personnel Cost	IDR	616,667	166,667	210,000	205,556	438,889	375,000	337,500	250,000
Water resources source		GW[132]	GW	GW	GW	GW	SW	GW	SW
Initial Consumers	House Connection	400	400	350	370	180	300	400	200
Current Consumers	House Connection	1500	620	882	688	873 (305)	460	927	259
Growth Rate	(%)	275	59	152	86	330	53	132	31

[131] Operating ratio: operating expense/net sales (similar to working ratio: total expenses (including financial costs)/gross income)
[132] GW: groundwater, SW: surface water (ie. River or pond)

Secondly, two somewhat paradoxical characteristics of the consumers served by the HIPPAMs make the relatively low tariff important in terms of securing revenues. The first development is that Lamongan Regency has experienced a higher than average economic growth over the past years. This has led to an improved average household income (Statistics Bureau, 2015), which may have made it relatively easier to introduce water fees. At the same time, the percentage of consumers that are classified as belonging to the lowest-income bracket is between 50%-60% of the households (WSP, 2011b). This dual development of increasing income in an area which is relatively poor and where other acceptable (quality) alternatives are absent made the purchase of cheap water from HIPPAMs an attractive option.

HIPPAMs in Lamongan try minimizing commercial risk by how they decide about service extensions. When a potential customer requests a service extension, the HIPPAM will conduct a cost-benefit analysis prior to agreeing to the development of this extension. Based on an assessment of the capital and operational costs involved and the potential revenues a new connection(s) will generate, the HIPPAM will decide about the service extension. Here it is important to note that HIPPAMs, as alternative providers, do not have an official mandate to ensure universal service coverage. This means that they can, and in fact do, reject requests for service expansion if they consider the expansion and subsequent operation to be commercially unviable. A staff member of one of the HIPPAMs reported that "*service coverage almost reaches 93% for this village. There was demand from another village but I thought it was not feasible due to the (long) distances and that village's characteristic that it has spread out houses leading to a high cost per connection. This was not advantageous for the HIPPAM*"[133]. Hence, HIPPAMs may consciously decide not to serve certain segments of the population when it endangers their financial viability.

During the initial project implementation phase, the eight CBOs were trained by either World Bank or AusAid consultants to incorporate management practices that would foster efficient operations. In these trainings, the consultants initially promoted the idea of direct punishment and sanctions as a way to promote billing-collection rates: consumers would be cut off from the service in case they delayed or failed to pay their bills. However, HIPPAMs were reluctant to implement this in the closely knit communities in Lamongan Regency[134]. Most HIPPAM leaders considered disconnection measures "difficult and not suitable", as many of the village

---

[133] HPM3A (December 2016)
[134] HPM3B (December 2016)

residents are related to one another through employment or kin[135]. They instead preferred embedding their water services in already existing and well-established social institutions, such as the local village committee or the community mosque. It was through these institutions that consumers would experience the pressure or moral obligation to pay for their water services. In fact, most consumers would not only feel obliged to pay, they would also tend to pay on time. For example, when the request of paying for services comes from the village chief who also represents the HIPPAM "consumers tend to be afraid not to pay"[136]. One of the CBOs studied appointed the leader of the local mosque, a highly respected person in the community, as the Head of the CBO. In another case, the Head of the HIPPAM was a village elder who enjoyed considerable respect within the village. As explained by one of the consumers: "We trust the CBO leader because of his active contribution to the village's development". The presence of village leaders in the CBOs appears to generate a sense of trust and legitimacy among consumers, which makes operational and financial decisions to be readily accepted. "The Head [of the CBO] is usually a person trusted by the community and they have a good ability to communicate with the community including the delivery of HIPPAM's 'message'"[137].

Indeed, an important factor in the HIPPAMs successful existence is their ability to integrate their operations in village life and institutions, beyond water provision. Originally designed and implemented as water service providers for the villages in Lamongan Regency, many HIPPAMS added other societal purposes to the organization during their development. By thus engaging in the social and political life of the villages, the HIPPAMs enhanced their contribution to the community (locally known as *iuran*). In this way, rather than just being a water provider, HIPPAMs have become and manifest themselves as 'village development agents'. HIPPAMs have for instance financially supported the development of local roads, a sewerage system, street lightning and even urban gardens. By moving beyond the boundaries of water service provision, HIPPAMs wield acceptance and support for the organization among the members of the community. This, in turn, also helps in collecting bills for the water provided.

[135] HPM4A (December 2016)
[136] HPM3A (December 2016)
[137] HPM8A (November 2016)

a. Reducing capital and operational costs

To be commercially viable implies that HIPPAMs make sure that their stream of income actually covers the costs. Personnel-related costs are generally among the higher expenditures of water service provision. In the case of the eight HIPPAMS studied in Lamongan, however, the average monthly cost for staff is low. It ranged from less than IDR 200,000 to about IDR 600,000. This is very low when for instance comparing it to the minimum monthly salary, set by the Lamongan Regency Government, which is set at IDR 1.7 million. HIPPAMs can keep salary costs so low because the committee members and staff usually have a different primary profession such as teacher, public officer or farmer. As one HIPPAM leader explained *"the salary I earn from my profession as teacher has been sufficient to finance my family life, therefore I am not looking for money here in the HIPPAM"*[138]. Rather, for most of the members of these committees their commitment to the functioning of the HIPPAM is of a 'social nature'. The good image of the work the HIPPAM does for the community played an important and positive role in explaining their willingness to work for the organization. By paying such low average salaries, the personnel costs of the HIPPAM are generally below 30% of their total revenues (Figure 7).

Given that all eight HIPPAMs have operating ratios below 1.0[139], this means that the percentage of personnel costs as part of the total costs structure is even lower than the percentages mentioned. This implies that the rest of the 70% to 89% of the revenues can be used to finance other costs or initiatives.

Figure 7 Personnel Cost and Personnel Cost as a Percentage of Revenues (Source: Surya, 2017)

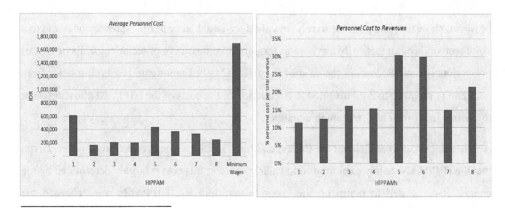

---

[138] HPM1A (November 2016)
[139] Operating ratio shows the capacity of a company to cover operating expenses (thus, not debts) with revenue generated from sales. The lower the ratio, the greater an organization is able to generate profit.

In comparison, a World Bank (2017) study analyzing the aggregation of services in small towns suggests that staff costs often make up 60% of the costs for small-scale water providers. Even large scale water utilities, which are able to recap economies of scale, often have levels of personal–related costs which are in the vicinity of 30%-35% (Tynan and Kindom, 2002).

That the HIPPAMs in Lamongan succeed in keeping personnel costs so low is strongly linked to the CBO´s embeddedness in existing social institutions in the village. As noted, HIPPAMs are mostly staffed by people that have another primary source of income. The incentive to work for the HIPPAM committee lies not in the money it earns, but in the pride and prestige a position in the committee brings. The trust and power that others have in members of the committee helps individual members to gain respect and possibly advance their political ambitions. One member of a HIPPAM committee explained how his position in the HIPPAM had helped him run for the office of Village Chief, ultimately defeating the incumbent candidate[140]. Most committee members take the work they do for the HIPPAM very seriously. As one committee member mentioned jokingly: "*I work as an education supervisor for a pre-school. However colleagues labelled me as 'water supervisor' because I often prioritize the HIPPAM*"[141].They generally consider the position they hold in the HIPPAM committee a 'dedication to the village community', rather than as a way to pursue additional income. Their commitment and seriousness contradicts a widespread belief in the policy literature that volunteer-based labor is a barrier to professionalization and stands in the way of commercialization (see World Bank, 2017; Moriarty et al.; 2002). Indeed, resorting to volunteer-based labor has been identified by many as a factor preventing CBOs from operating 'professionally', as volunteers would lack technical and managerial capacity as well as the motivation to operate the systems (Moriarty et al, 2002). The HIPPAMs in Lamongan Regency, however, show that it is possible to rely on volunteer-based labor to run water services and that volunteerism is not necessarily a synonym for poor administrative or technical skills or lack of professional ethos. Rather, in the case of the HIPPAMs in Lamongan, resourcing labor from community members has contributed to a reduction of the costs of the HIPPAM allowing them to operate in a financially sustainable manner.

Another important cost element are the expenditures related to the production of water. Given that the HIPPAMs mainly exploit groundwater resources, this cost is largely determined by the energy costs needed to pump up the groundwater. Initially, HIPPAMs were charged the

---

[140] HPM7 (November 2016)
[141] HPM1A (November 2016)

electricity tariff for industrial use. Their consumption of over 3,5000 Kwh/month, required to provide water services, positioned HIPPAMs in the category of industrial users. With the help of village leaders and the local government, however, the HIPPAM committees successfully negotiated to become eligible for the social tariff with the electricity company (PLN)[142]. They succeeded in being considered as a social organization as "*there is no rule that categorized HIPPAM as industry, so based on the nature of the HIPPAM we think we can get the social tariff for the electricity*"[143]. Such a social tariff could translate into a 40% reduction of total electricity costs for the HIPPAMs[144].

A significant feature of the HIPPAMs in the Lamongan Regency is how they are regulated. HIPPAMs are currently only regulated by INGUB 1985, which is the Governor Instruction of 1985. This regulation recognizes the possible existence of HIPPAMs, but does not reflect the relevance that HIPPAMs currently play in the Regency in providing water services[145]. This condition has both advantages and disadvantages for the HIPPAMs. On the one hand, it places HIPPAMs in a fragile position as their status as water providers is largely unprotected. This is illustrated by cases where conflicts between HIPPAMs and PDAMS occurred. Illustrative is the expansion of the HIPPAM in Sukomulyo (HIPPAM no. 5), which conflicted and overlapped with the service area of the local PDAM. The rights of the PDAM, which are owned by the local government, outweighed those of the HIPPAM. Such conflicts, in which the local government will side with the water enterprise of which they are the owner, can significantly reduce the revenues of the HIPPAM and worsen their financial situation. However, the 'lack' of regulation also allows the HIPPAMS considerable room for maneuver. Since HIPPAMs are not registered as official water operators they are free to define with which technological approaches they choose to serve their customers. They are not obliged by any law to comply with technical standards and their water quality is not controlled. This has obvious impacts on the water quality they provide and may negatively affect the final user. The policies concerning HIPPAMs in Lamongan are implemented and interpreted through informal communications between governmental actors and HIPPAMs rather than through legislation. Or as explained by a representative of the HIPPAMS Association "*...the culture has already been built, but the structure [regulation] is not*"[146]. The local government, as the authority responsible for the

---

[142] *PLN (Perusahaan Listrik Negara) is a State Owned Company responsible Electricity Supply throughout Republic of Indonesia*
[143] BA1 (November 2016)
[144] HPM2A (November 2016)
[145] PG1 (November 2016)
[146] BA1 (November 2016)

supervision of the water quality provided by the HIPPAMs, also does not attempt to make HIPPAMs improve their water quality. A local government officer explains that "*we understand that we cannot force them to comply with the water quality standard, therefore we just 'close our eyes' when it comes to this issue*"[147]. The lack of regulation allows the HIPPAMs in Lamongan to reduce their costs and offer cheaper services as they support a less complex technological water supply system. If they would have to comply with quality standards, their operational costs would increase and it is highly likely that HIPPAMs would then have to increase their tariffs.

3. Discussion

In most policy-oriented literature, the commercialization of CBOs providing water services is advocated based on the assumption that commercialization and associated practices will lead to more sustainable organizations. For HIPPAMs in Lamongan Regency, however, the situation is somewhat different. It is not so much the strict operationalization of commercialization that drives HIPPAMs to operate on a financially sustainable basis, but rather the fact that HIPPAMs have had no choice but to pursue financial sustainability given that they have largely been ignored and neglected by local government. Although commercial principles were instilled in the original CBO model by external donors, for the HIPPAMs cost recovery is a pure necessity to survive as they cannot rely on the local government for additional funding should the need arise. In turn, the fact that they do not have to financially support CBOs is precisely what makes the commercial CBO model attractive for financially-strapped local governments, as it helps them keep their expenditures low (WSP, 2011).

To make CBOs commercially viable, recent policy-oriented literature argues that "[p]rofessionalization of community management with policy embedding, adequate legal frameworks [and a] move away from voluntarism" is a key building block for improving rural water services delivery[148] (World Bank, 2017:15). The ideal community based organization appears to be one where a "water system is operated by an authorized business-like organization with a community institution either taking responsibility for service provision in a professional way or outsourcing this to other entities" (Hutchings et al., 2015:969; see also Moriarty el al, 2012). Emphasis is thus placed on professionalism and the business-like nature

---

[147] HPM2A (November 2016)

[148] In addition to the professionalization-approach for CBOs, a 'classic' approach exists. In this approach improved functioning of CBOs is envisaged on the basis of "extensive, long-term support" (Hutchings et al., 2015: 980). However, this approach would require continued support from external agencies, which the HIPPAMs do not have access to.

of the CBO. In looking at CBOs in the Indonesia Al'Afghani et al. (2015:103) link professionalization to the de-politicization of the CBO. They argue that "[i]n terms of professionalization, it would be ideal if the technical apparatus is insulated from local politics…". This would imply that "the head of the village should not be in a management position but in the supervisory board, to prevent this conflict within the CBO daily operation" (Al'Afghani et al., 2015: 103). On paper, these ideas have guided the establishment of the CBOs in Lamongan Regency. The principles underpinning projects, such as WSLIC and PASIMAS, focused on the need to capacitate CBOs through trainings, support and incentive programs that would gear their operations towards market-inspired practices. They envisioned these community organizations as viable independent providers. The trainings therefore focused on financial management and enhanced billing collection. 'Loans' were sometimes provided to incentivize efficient practices, as it would force the CBOs to engage in activities that would generate enough revenues to repay the loans. Indeed, the underlying principles on which the projects WSSLIC and PAMSIMAS were based were translated into characteristics of a typically '"business-like organization" operating in a "professional way" (Hutchings et al., 2015:969). In spite of all the attempts to instill more professionality and a business spirit in members of the newly established CBOs, the commercial success of the eight HIPPAMS researched in this thesis is not the result of their professional and business-like character, or their strict following of the principles. Instead, the key to success appears to lie in how the HIPPAMs have embedded the organization of water service provision in existing socio-political institutions. They for instance appointed village leaders as members of the management committee of the HIPPAMs, or used the profits from operations to fund village development in the form of roads or schools. It seems fair to say that contrary to the argument that water providers should operate at arm's length of the political domain (Baeitti et al., 2006; AlÁfghani et al., 2015), it is actually the active politicization of the HIPPAMs at village level which has allowed them to achieve high billing-collection rates. This engagement with the political domain of the villages and the benefits derived from it is very context specific (Rizza, 2006). The interest of village leaders in the HIPPAM is that it provides a platform that grants them prestige, with volunteering work also being considered as fulfilling a moral duty.

What characterizes the HIPPAMs is a combination of different value systems. Their functioning in fact blurs the distinction between markets and civil society that guides much thinking in the water sector (Evers, 2012, as cited in Jager and Schroer, 2014). The HIPPAMs not only actively maintain and operate according to community values, but they also mobilize these community values to achieve financial sustainability. In this sense, HIPPAMs can be characterized as hybrid organizations. Policy thinking has it that hybrid organizations are doomed to face incompatible interests, as they need to respond to different logics and demands (Hes et al., 2014). In the case of the HIPPAMs, it is precisely their ability to tap from different normative repertoires that enables them to integrate the institutions of civil society and those of the market. Rather than achieving their optimal operational performance by disaggregating and becoming disconnected from their cultural and political environment as literature on autonomy suggests (Braadbaart et al., 1999), HIPPAMs perform successfully in terms of cost recovery and commercial viability precisely because of how they have integrated themselves in existing community institutions.

Although the CBO model as implemented in Lamongan Regency appears financially sustainable, this sustainability does not necessarily translate into good, or even adequate, services for all consumers. In fact some of the elements which explain financial success, may lead to the reproduction and strengthening of existing inequalities and inadequate services (see also Rusca and Schwartz, 2018). First, although the demand conditions ensure a significant demand for water services and the organizational structure allows the HIPPAM to provide these services at a relatively low tariff, the HIPPAMs have also opted to limit the treatment process of water provided. As a result, water of sub-standard quality is provided to consumers. The favorable demand conditions are useful for the HIPPAM from a financial perspective, but from a consumer perspective, the lack of affordable alternatives to water services also means that they are captured by a single supplier. Secondly, the existing social institutions used by the HIPPAM have been beneficial in a financial sense. These have led to a relatively high billing-collection ratio and also allowed for low personnel costs. This research did not include the perspective of the users, but it is quite possible that the choices made by the CBOs and the political control exerted by them has negative implications (Rusca and Schwartz, 2018; Cleaver, 2010) in terms of potentially depriving some consumers from securing access to water. In the Lamongan Regency, practices were identified that prioritize financial sustainability over universal service coverage. The HIPPAMs have significant discretion in choosing where the systems will be expanded to and which areas may be unable to access water.

The relatively low level of services (i.e. water quality) received by consumers of the HIPPAMs also suggests the existence of an apparent paradox or discrepancy between the ideal model and the practice of commercialization. In the ideal model of commercialization, the water provider acts independently of the government and operates on the basis of cost-recovery. This is argued to be a desirable situation as it means that the water operator is mainly accountable to the consumers it serves, as the revenue required to cover costs must be collected from these consumers. The model of commercial water provision thus emphasizes consumer-orientation (Baietti et al., 2006; Schwartz, 2006). In the practice of commercialization as implemented by the HIPPAMs, however, the demand conditions prevailing in Lamongan regency allow the HIPPAMs to reduce costs by providing relatively low services. The fact that consumers in Lamongan regency lack alternatives to the HIPPAMs for accessing water means they are essentially captured consumers. This ´captivity´ reduces the need for the HIPPAM to be accountable to the consumer. I therefore conclude that consumer-orientation and commercialization as it is implemented in the context of Lamongan Regency may be less strongly linked than much of the literature on commercialization would suggest.

It is relatively simple to understand why the HIPPAMs are viewed as a success by donors and the Indonesian (local) government. The HIPPAMs have been useful for the donor to promote their project. They have also been useful for the government, which is looking for 'solutions' to their 'overburdened' public system. Finally, it is useful for the HIPPAM and village leaders as jobs are generated and water is delivered. However, seeing the CBOs as a (commercial) success without taking into account how the strategies for commercialization have impacted different categories of consumers reveals only one side of the story. Moreover, promoting the commercial success of the CBOs without acknowledging the complex and context-specific mechanisms that contributed to delivering this success does not contribute to improving the understanding of commercialization, or of organizing water provisioning for that matter.

Chapter 8: Talking and Practicing Commercialization

I started this thesis, in chapter 2, by explaining how a coalition of international agencies, donors and (national) governments have embraced the principles of commercialization as the key to improving water service provisioning. Commercialization has pervaded the provision of water services to such an extent that the principles of cost recovery and of political autonomy have become important markers of good performance. Hence, progress in water service provisioning is importantly measured by the degree to which utilities are able to recover costs, and by their ability to operate independently from the government. In this concluding chapter, I use the empirical findings of chapters 5, 6 and 7 – documenting how operators implement commercialization in practice – for a reflection on the 'success' of commercialization. In the chapters, I showed that actual water provisioning in the small towns in Mozambique, Uganda and Indonesia deviates in significant ways from the ideal of commercialization, challenging whether these ideals (and the principles of cost recovery and political independence) are at all realizable. These deviations happen with the active support and endorsement of government agencies and funders. I also showed that pressures to recover costs and be financially sustainable sometimes force operators to sacrifice quality of service or to postpone the extension of services to remote or low-income areas. This calls into question whether commercialization can and will lead to the improvement of services, and whether it will ever contribute to reaching the goal of making water accessible to all. In spite of the deviations and disappointing outcomes, the belief that commercialization will improve water provisioning is still strong. The trust in these principles is actively upheld by donors and government agencies, but also by the operators themselves whose practices challenge it. After having summarized the main findings of the previous chapters, I end this chapter by reflecting on why this paradigm for water service provision remains unchallenged. I also provide some ideas about ways to improve policy learning. I propose that a focus on the practices of operators can provide new inspirations for thinking and doing water provisioning, and perhaps thereby helping meet the challenge of making water accessible to all.

2. Practicing   commercialization: deviations   during implementation

As explained in chapter 4, I use a conceptualization of policy implementation that considers implementation as an intrinsic part of the policy process, and those that implement policy models as active policy actors. This conceptualization of policy implementation hinges on the acknowledgment that for policies to shape actual practices and outcomes, they need to be translated into operational realities. It also means, that translation will necessarily involve deviations and modifications from the model or the ideal, as the realities in which the policies are implemented are different. For understanding the commercialization of water provision differently, therefore, it is important to not just measure outcomes – i.e. rates of cost recovery; numbers of people served; non-revenue water or billing collection rates etc. – but to also understand how these outcomes were achieved.

The three cases of everyday operating practices of providers in Mozambique, Uganda and Indonesia presented in the previous chapters show that for utilities to be seen as operating in a commercial way, they need to (1) strive to achieve full cost recovery. They can do this by having consumers pay for the water they receive, something that requires tariffs to reflect the real costs of providing the service, or reducing their expenses; and (2) they need to operate at arms-length of the government, which would lead to operational choices, which are not subject to political reasoning, but rather centered on delivering services of better quality for consumers. Below, I summarize what the operators in the studied cases did to meet these requirements.

Re. 1: **Full cost recovery.** The service providers studied in the small towns in Uganda, Indonesia and Mozambique all engaged in serious efforts to achieve some form of financial sustainability and cost recovery, and all succeeded in showing positive results. One important way they did this was by reducing their costs, by:

— *Avoiding or reducing water treatment (costs)*: In the case of the HIPPAMs in Lamongan Regency in Indonesia, the operators decided to not treat the water prior to its distribution to consumers, because they assumed that the groundwater was of good enough quality. Similarly, in Bushenyi-Ishaka the operator simplified the infrastructure for water production to reduce the operational costs related to treatment. Their justification is that the systems exploiting springs require less treatment, because they considered the water of the spring to be of 'sufficient' quality.

— *Foregoing or delaying network maintenance.* In Uganda, the operators reduced costs by minimizing repairs and maintenance, or by delaying maintenance work. In this way, they keep the system running or marginally improve it without having to spend much. As highlighted in the next section, the providers prefer investing in extensions of the system rather than in the maintenance and upgrading of the existing system.

— *Economizing on material costs.* In Mozambique, the asset holder (AIAS) provides a degree of support to develop major construction works and expansions of the distribution networks in the form of direct construction and replacement of equipment or the provision of water meters to accelerate connection rates. In order to reduce investment costs the materials used for the expansion of the network, at times, do not fully comply with the standards stipulated to guarantee safe delivery of water to the final user. Similarly, the HIPPAMs in Lamongan Regency use very basic infrastructure for the delivery of services.

— *Economizing on labor costs.* In Indonesia, operators source free labor, through community engagement, for the operations of the provider. This labor is used for activities varying from technical support to bill collection. People in communities are willing to volunteer or work for lower salaries in exchange for the social and political benefits that their participation in the HIPPAMs provides. In Uganda, NWSC organizes its Branches in such a way that a cluster of systems can make use of one or two specialized employees. For example, a cluster of Branches would make use of one quality assurance engineer, allowing NWSC to save labor costs.

What is noteworthy about these cost reductions is that they operationalize the idea of commercial viability in very creative ways, in the process re-defining and even challenging the very idea or possibility of cost-recovering tariffs. Operators achieve cost-recovery not so much by reducing redundancies or trimming the fat off the organization (as the commercial principles of the original model would have it), but importantly by transferring (to others or to the future) or altogether avoiding some of the usual expenses that providing water of a reasonable quality entails.

In addition to reducing costs, the operators in small towns also try to meet expectations of commercial or financial sustainability by increasing their revenues. In the policy prescription, tariffs that match or reflect the real cost of producing and distributing drinking water are a key element to increase revenues. Yet, for none of the studied water operators, raising tariffs was a feasible or do-able strategy to increase their income. In Mozambique, regulators firmly controlled the tariff. For operators, this made the process to amend tariffs slow and highly bureaucratic, thus effectively discouraging them to do so. Uganda uses a uniform tariff across the country, meaning that the operator in the small towns of Bushenyi and Ishaka had to adhere to the same tariff as the one used by operators in larger urban centers such as Kampala and Jinja. For Indonesia, even though the HIPPAMs would be allowed to charge higher tariffs, they were reluctant to do so as they feared that customers would not be able to pay those and would even resist them. In all, the cases show that the political and institutional context in which providers operate gives them very little powers to set a tariff that is high enough to increase their revenue and allow for cost-recovery.

Apart from lacking the powers and ability to set tariffs, water providers, such as the HIPPAMS, were also hesitant to raise tariffs as they realized that it would influence the quantity of water sold to consumers, especially if alternative sources of water[149] are available. The existence of such alternative sources means that consumers may opt to diversify their water sources, choosing different sources for different uses depending on the (perceived) quality of these different waters. As the quantity of water consumed for drinking and cooking is relatively small in comparison to that of water used for other purposes like cleaning, higher tariffs may provide an incentive to the consumer to reduce the use of relatively expensive water from the piped network[150]. Hence, when alternative water sources are available, increasing the price of water may prompt consumers to buy less water as they can source it more cheaply from elsewhere instead. For the operators, this means that the time needed to recover investments will increase. This does not just explain the operators' hesitance to raise tariffs, but also why they are reluctant to expand services to areas with alternative water sources[151]. Uganda is a clear case in point. Here, the presence of water from open springs at no cost for the consumer negatively impacted their willingness to connect to the piped network of NWSC. Moreover, even if a consumer was connected, the availability of free alternative sources limited the amount of water they consumed from NWSC. The Lamongan Regency Indonesia is the only of the three

---

[149] Alternative sources may include springs or rivers that consumers can easily access or other delivery options, such a municipal standpipes or shallow wells that allow access to water at lower prices.

[150] One way in which water operators have tried to stimulate water consumption from the piped network is by increasing the fixed part of the tariff (in relation to the volumetric part). By having a relatively large fixed part of the tariff, the impact of the volume of water consumed on the overall bill becomes less.

[151] Although alternative sources are also used in urban settings, the rural settings of small towns make the accessibility of such water sources easier. This phenomenon is not limited to developing countries. The availability of alternative sources was also a challenge for the expansion of water companies in rural areas of the Netherlands (Blokland, 1999). This challenge was partially addressed by reducing the volumetric component of the tariff, meaning that these tariffs had a relatively large fixed part, so that consumption levels did not greatly impact the water bill for the consumer (Blokland, 1999).

cases where there were no cheaper and accessible alternative sources of water. This is what allowed the HIPPAMs to expand services: consumers had few opportunities to opt for other water sources.

In all cases, the preferred strategy of operators to increase their income is the expansion of the pool of users by extending the network. The cases show how operators did this very selectively, so as to ensure that the investment costs of network extensions would be easily and quickly recoverable. As noted, in both Uganda and Mozambique, infrastructural investments for new connections were prioritized over investments in maintenance or improvements in the distribution network. This is because such prioritization more directly benefits the revenue generation capacity of the operator. Operators choose where to extend on the basis of 'economic sense': those potential consumers who are considered too expensive to connect or who are not expected to generate sufficient revenue are not selected for service expansion. This was most apparent in Uganda, where the initial expansion of the network of small towns acquired by NWSC largely adhered to the logic of first connecting those areas that were deemed most profitable, either because connections were guaranteed or because higher population densities were expected to lead to a higher rate of connections per kilometer of pipe.

Re. 2. **Independence from government agencies.** A cornerstone of the commercial water services models is that operators should be autonomous: they should be financially and politically independent from governmental agencies. As such, these models envision that water operators are either completely independent (private sector organizations) or that they are public organizations that operate as independent entities at arm's-length of the government. In Mozambique, the delegated management model relies on two simple but very important premises for succeeding in making water operators autonomous. First, there needs to be a well-developed private sector, with water operating businesses ready to engage in contracts with AIAS. For the model to work, these private operators need to be knowledgeable and able to independently operate water systems. Apart from this operational knowledge they also need to have the commercial expertise to identify and develop opportunities for enhancing their business. Second, there should be enough operators to choose from as realizing the advantages derived from delegated management requires some degree of market competition. In theory, the asset holder can then choose from a pool of qualified private operators to select the one that is most likely to offer the best services. Neither of these two premises existed in Mozambique when the model was adopted[152]. The asset holder (AIAS) therefore had to be creative in operationalizing the model. It did this by reinterpreting its role from being only an asset holder, or a leaser of infrastructure, to a role, which combined asset holding with a role of broker of capacity development projects to promote and create private operators. Rather than simply delegating responsibilities to the private sector, AIAS took on an active role in supporting a still embryonic private sector with the aim of developing it further. In doing so, AIAS did not operate at an arm's-length distance from the operators, but actively engaged with and interfered in them.

Whereas the delegated management model in Mozambique represents a case in which an external government agency is actively involved in supporting service providers, HIPPAMS in Indonesia never were strongly regulated. Here, the supporting role that the model envisioned for the local government never materialized. In this respect, HIPPAMs appear to conform to the ideal promoted by the policy model: they are relatively isolated, and thus autonomous, service providers. Yet, for HIPPAMs to function in a viable manner, they made a deliberate choice to deviate from the model in a different way: they expanded their activities beyond just water, interweaving the provision of water services with active engagement in local government matters. In this sense, the HIPPAMs became broader development agencies in the villages they serve. By making local (political and religious) leaders active members of their organization, they obtained legitimacy and authority. The HIPPAMs thus purposively embedded water provision in broader government and community institutions as a way to ensure their ability to collect fees. In this way, the HIPPAMs themselves became (part of) local government and thereby defied the model.

What is striking in the three studied cases of the implementation of commercialization is that operators continued to rely on external forms of support to subsidize their operational costs or finance new investments. They mobilized this support either from the government, sometimes from donors, or from the community itself. In Indonesia, operators make use of their volunteer-base to reduce their staff related expenditures and engage with government institutions to ensure high levels of bill collection. In Uganda and Mozambique, the private operators or branch managers reduce their direct costs by recruiting outside support (in the form of cash or

---

[152] The absence of a pool of experienced local private operators makes the choice for delegated management in Mozambican small towns an interesting one. Most organizations promoting the delegated management model do so on the basis of the belief that the private sector not only has operational expertise and knowledge, but also that they have incentives to use that expertise to efficiently provide services. Without the existence of a private water sector this basis for choosing the delegated management model largely disappears.

equipment) for the expansion of services, or the maintenance of the infrastructure. In Uganda, NWSC continues to receive funds from the Government of Uganda to develop water infrastructure across the country. While NWSC claims to use internally generated funds to develop infrastructure in small towns, in practice the Ministry of Environment and Water transfers funds to the small town operators to help them develop infrastructure. This dependency on funds influences operational decisions and choices concerning the nature and location of water infrastructure development. Illustrative of this influence is the accelerated takeover of towns over the last years by NWSC. If it was solely up to the NWSC they would have spread out the takeover of small town systems over a longer period of time, allowing them to initially concentrate investment in the most commercially viable small towns. However, because they have strong interests to remain on good terms with the Government (through the Ministry) they accept the accelerated takeover of small towns. The importance of the connection with the Government of Uganda thus outweighs the immediate operational interests of the utility.

What these findings suggest is something of a paradox concerning the two principles of commercialization. It appears that the way operators strive for commercial viability and cost-recovery can only be understood by recognizing that they are financially and politically related to local and national political actors. The cases suggest that the separation of politics and administration preached by international policy recipes is unlikely to happen: the ideal of substituting political control with managerial control in order to provide more effective services is not a realistic one, at least not for the small towns studied (Hefetz and Warner, 2012: 293).

3. Does commercialization improve water provisioning?

Official accounts and studies of commercialization generally show impressive increases in cost recovery and financial sustainability, lower levels of non-revenue water and increased coverage. Hence, such accounts confirm that commercialization 'works'. How these successes are achieved is, however, seldom examined in much detail: most studies use inductive types of analyses that compare the situation before the introduction of commercial principles with the one a few years after. I discuss this in more detail in the next section, suggesting that this happens because the actors involved in the promotion and/or implementation of *commercialization* have an interest in continuously reiterating its success. Here, I want to summarize the findings of our studies to argue that the idea that commercialization – in its pure form – will help provide access to water services to all needs to be questioned.

First of all, our cases of commercialized water provisioning in small towns show that the fear of 'cherrypicking' – often used by those critical of privatization and commercialization – is a valid concern. The pressure to be financially viable and recover costs pushes operators to prioritize investments to those areas and customers that are likely to generate fast and highest profits and revenues. For those not connected, their 'strategic attractiveness' plays a very important role in determining whether they will be able to connect to the network. This will depend either on their geographical location in the town, or their (perceived) economic capacity to connect and pay for water. Extending services enlarges the market supplied by the service providers. Both in Mozambique and Indonesia the providers have strategically expanded their services to those areas that are considered to be most profitable, either because of potential for revenue generation was promising or because the costs associated with expanding services to a specific area would allow for a large number of households to be connected[153]. If the provider is not convinced that establishing a new connection can be done in a commercially viable manner they will opt to forego establishing this connection.

Also, for those already connected to the network, the push for service expansion to unconnected households may negatively impact the quality of the service they receive as maintenance or upgrading of the existing system is forfeited to allow for the provider to connect additional consumers. In order to improve their chances to operate on a commercially viable basis, operators may also compromise on the rehabilitation of infrastructure and maintenance by using sub-standard materials. This may result in lower levels of services in terms of the quality of water delivered and the continuity of service provision. The cases studied thus suggest that there is a trade-off between the pursuit of commercial viability and the level of service that the consumer receives.

Whereas most literature on commercialization suggests that tariffs will be set at levels to allow for a sufficient quality of services, the cases studied in this thesis show that the particular interpretation of commercialization by the operators in small towns (reduce costs and selectively increase the market) and the agendas of national governments and donors may cause commercialization to lead to lower level services for existing consumers and the exclusion of some households from accessing the network altogether. That the quality of service for existing consumers may decrease also reveals the existence of an apparent paradox in the underlying reasoning of the ideal model of commercialization and the practice of commercialization. In

---

[153] One of the indicators that a water provider is interested in is the number of connections that can be established per kilometer of network. More connections per kilometer reflect of higher density of connections.

the ideal model, the water operator acts independently of government and operates on the basis of cost-recovery. This is argued to be a desirable situation as it means that the water operator is primarily accountable to the consumers it serves: after all, the revenue required to recover costs must be collected from the consumers. The models of commercial water provision thus emphasize consumer-orientation (Baietti et al., 2006; Schwartz, 2006). In the practice of commercialization, however, the water operators need to ensure that the agendas of the donors and national government are also addressed. This means that they need to adopt the targets of these agendas as part of their performance objectives: operators also (or perhaps largely) are or become accountable to the government and donors. The practice of commercialization thus arguably leads to less consumer-orientation than the ideal model would suggest.

4. The persistence of the commercial ideal

As highlighted in the previous section, small town operators' practical strategies and practices to (appear to) meet objectives of commercial viability deviate from what policy models promote and prescribe. Yet, their adjustments and deviations from the policy model do not seem to affect the belief in the model: the policy ideal of the commercialized water operator remains very much alive in contemporary policy discourse and in donor projects. In this section, I reflect on why this is so.

One way of explaining the persistence of commercialization as the ideal model for water service provisioning is to look at what policy models do, next to providing structure or guidance to the organization of service delivery. The three dominant models involving commercialization also represent a discursive agreement between different actors (water organizations, donors, government agencies) who depend on each other for realizing improvements in water service provision. These actors who each have different interests and who are linked to different domains - the international policy domain, the national policy domain and the domain of operators and consumers - need to agree on how to best achieve the goals of providing access to water to all. In this sense, commercialization also serves as a language that allows comparison and commensuration of water service provisioning across cases and countries. In this language of commercialization financial sustainability, including full cost-recovery and autonomy, have become the undisputable bottom-line (Rusca and Schwartz, 2018) in organizing the provisioning of water services. Why this language is the language of commercialization is most likely related to a much wider enthusiasm for neo-liberalism in international policy circles, with a strong preference for service provision to be guided by (quasi-)market principles. The language and ideal of commercialization have become so all-pervading that they have gained "a life of its own" (Molle, 2008:131).

As the language of commercialization particularly relates to a discursive agreement between primary actors, the question whether water providers in small towns actually implement commercialization or will be able to do so in the future becomes less important. More important is the ability of actors to express themselves in the terms of the ideal, in this language, thereby also appearing to adhere to it. In the policy literature, this phenomenon has been referred to with the term 'magic concept' (Pollit and Hupe, 2011). A 'magic concept' is "broad, normatively charged and lays claims to universal or near universal application" (Pollit and Hupe, 2011: 643). Magic concepts are broad enough to allow for multiple interpretations. They promise bringing about new and more 'modern' futures, thereby directing contentious political questions towards a not-yet attained tomorrow. This capacity of obscuring or denying conflicting interests and logics makes them ideal for use in official policy documents and deliberations. They remain alive and continue to circulate not because they actually help achieve desired outcomes, such as universal service coverage or financial sustainability, but because they provide a relatively easy way to arrive at a consensus, on agreeable terms.

The importance of such consensus becomes clear when realizing how the different actors involved in the negotiations about the organization of water services provisioning all depend on each other. In all cases studied, operators require (continued) external support to be able to continue operating and offering services in a manner that appears financially viable. Much of this monetary support comes from donor agencies. These donor agencies need to justify their own existence by demonstrating the success of their grants and projects to their constituencies. They do this by showing that their financial support translates into success, in the form of improved performance: more people with access to water services, for instance, or a reduction in non-revenue water. For example, Dutch Development Aid wants to contribute to the achievement of SDG target 6.1 by expanding sanitation services to 50 million people and expanding water services to 30 million people[154]. They need to continuously show that the projects they fund do contribute to reaching these targets. Similarly, national and local governments have an interest in ensuring service extension within their countries and municipalities: being able to show progress in this regard is politically important. All this means that operators, governments and donors are bound together in interlocking relations of

---

[154] https://www.dutchdevelopmentresults.nl/theme/water#water_sanitation-hygiene (accessed 5 August 2019)

dependency: operators need to show that they are meeting donor objectives to continue receiving support, while donors and governments need operators to perform according to their plans to satisfy their constituencies.

Mutual agreement on and continued adherence to the dominant models of commercial water provisioning is a mechanism to make this mutual dependency work for all primary parties, at least on paper. For the water operators, conspicuously adhering to the model is a way to show to donors that they do what is expected of them, and thus to show that donor support was impactful. Somewhat ironically, they thus need to show they are financially and politically independent, to remain eligible for future donor funding. For donors and national governments, there is clear merit in showcasing the success of the model that they promote as it provides evidence of the effectiveness of their projects, policies and investments. Hence, all actors (the water organizations, the donors and the government ministries) have an interest in performing, that is actively displaying, the success of the model that is in vogue.

This also means that for the actors involved in these relations of dependency it may be risky and costly to reveal how the day-to-day choices and behaviors of operators deviate from or even challenge the model. Hence, normal reporting and accounting practices mostly use deductive methods: they use a number of key indicators to report progress and outcomes and assume that these were achieved through the cause-effect relations that the model hypothesizes. Hence, if water is (reported as) paid for, if operators can be shown to operate as independent entities that have staff with technical knowledge, and if software and reporting systems are in place, the models are considered to work. In short, normal evaluations will conclude that if inputs are provided for, the system will result in the desired outputs. That the cause-effect relationship between input and output is different from what the model suggests is not interesting as it would require opening up the model for questioning, risking to damage the reputation of involved actors.

What also matters in this context is that those evaluating the success of the model are often the same actors that are in charge of promoting and spreading the validity of these models through funded programs. None of the involved actors have an interest in questioning the model or to find ways of reducing the divergence between the ideal and the actual implementation. Rather, all involved benefit from showing that the models are successful, or achievable. The adjustments operators make to the models in their everyday implementation are therefore not interpreted as issues questioning the very principles of the model, but as correctable obstacles on the road to full success. Water operators who are dependent on financial resources and support from external donor agencies or government organizations need to consciously distinguish between their talk (portrayal) and their actions (operations). Despite the challenges in complying with the principles presented by the models of commercialization, these water providers make conscious efforts to portray an image of their operations that mimics the original model. Hence, for the service provider, questions of cost-recovery are less about which costs are exactly covered and to what extent these costs are covered by internally generated income and more about being able to show that water operators operate on the basis of full cost-recovery. The CEOs of NWSC have aptly illustrated this. Muhairwe may have realized that "cost-recovery is a myth" (Schiffler, 2015), but the organization continues to present itself as operating on the basis of cost recovery (NWSC, 2017).

In fact, in none of the studied cases were the operators able to achieve cost recovery by themselves: they continued to depend on outside forms of support. Yet, this did not prevent them from celebrating their commercial viability or claimed independence and autonomy. Operators have good reasons to structure their operations the way they do, and they play with local and national politics as much as they are subject to them. What is important in this analysis is the recognition of the need for these organizations to 'perform' the model as their existence is justified on the basis of this model. In the process of continuously having to perform the success of the model, some water utilities have come to consider a particular model as a strong part of their identity. Illustrative is the way that NWSC presents itself. Muhairwe's (2009) presentation of NWSC as "a successful example of a commercial public utility that combines public sector control with private sector efficiency" illustrates the degree to which NWSC has invested in becoming almost synonymous with the model. This is so even when the extension of services to small towns under the auspices of NWSC directly challenges their capacity to perform as a commercial enterprise. Indeed, NWSC's 'success' as a commercial public utility partially explains why its mandate has been expanded to cover small towns as well. The commercial utility, able to act autonomously and on a full-cost recovery basis is who NWSC is beyond the walls of NWSC. The need to perform the role of a commercial public utility is for NWSC but also the Ministry supporting the organization almost as important, if not more important, than the actual operational performance of the water provider. For organizations like AIAS and NWSC, which are heavily dependent on donor funding, becoming the embodiment of a donor-approved model allows them to access the funding they require. When the organization has become the representation of the model, and the model is based on the existence of that organization, much needs to happen to question the workings of that organization or model.

5. Is the divergence of talk and action a problem?

The interests involved in perpetuating the principles of commercialization in the water services sector force these utilities into a strategy of organized hypocrisy (Brunsson, 1993). As argued by Brunsson (1993:9): 'Hypocrisy is often associated with morals, particularly with poor morals. It is of course possible to argue that hypocrisy is a bad thing, which ought to be abolished at any price. There are however many arguments against such as view. One argument is that hypocrisy appears to be exactly what we demand of modern organizations: if we expose organizations to conflicting demands and norms, and expect that they should respond to them, then we must also expect hypocrisy". As Meyer and Rowan (1977: 352) have argued: "organizational success depends on factors other than efficient coordination and control of productive activities. Independent of their productive efficiency, organizations which exist in highly elaborated institutional environments and succeed in becoming isomorphic with these environments gain the legitimacy and resources needed to survive".

In the case studies, I have documented the many context-specific interpretations that operators make of cost-recovery and autonomy. Yet, the dominant policy and conceptual languages used to describe and make sense of the provisioning of water services force these messy and rich everyday details back into the neatness and predictability of the dominant policy models, making it seem as if actual water provisioning indeed confirms to these. As a result, actions and words develop in parallel realities. The effect of this is that operational realities – the changes, modifications, deviations to the model as presented and discussed in this thesis – rarely 'speak back' to the model, making it practically impossible to prompt a revisiting, questioning or challenging of the model's assumptions. Another way of saying this is that, despite the frequent routine interactions between actors at different levels, very little policy learning (change) is happening. While the different policy domains of operation, national governments and international donors, are interlinked, the influencing between domains only happens in one direction - from the policy ideal to the implementation. As explained earlier, the most important reason for this is that taking operators' practices more seriously would require reconsidering the underpinnings of the models. I conclude this thesis by arguing that a practice-based approach to the study of water service provisioning can provide a modest and pragmatic starting point for policy learning. One that can have a meaningful impact without necessarily overturning entire policy dogmas.

In the thesis, I have shown that attention to the empirical details of the deviations and modifications that operators make provides rich insights into what works, what does not, and why. Detailed attention to operational practices usefully enriches the conceptual vocabulary for talking about and discussing water services provision. This is important, first of all, because it makes it possible to refer to commercialization and water provisioning in non-totalizing ways. Currently, policy writings about the commercialization of water operators tend to take commercialization as the end rather than as a means to provide adequate and safe drinking water to all consumers. The degree to which commercialization is achieved is in itself a marker of performance. This makes it intrinsically impossible to imagine or acknowledge that a 'non-commercial' water operator works efficiently or effectively. Through the cases in this thesis, I have shown that there are multiple ways of embodying commercialization or of organizing water services provision. All the operators studied, in fact, succeed in providing continued services without fully complying with the ideal-type principles of commercialization. Moreover, the cases suggest that support through either money (through direct or cross-subsidies), material (to establish connections) or labor do not necessarily lead to inefficient practices.

Second, allowing for a different language to talk about commercialization and water service provisioning usefully opens up the range of possible ways in which operators achieve results. It allows also to separate practices from (types of) organization. Hence, the thesis shows that the categorization between private operators, commercialized CBOs or parastatal commercial entities is not necessarily a very useful one to understand how operators are working in practice. Instead of a concern with organizations being commercial or not, or financially sustainable or not, a vocabulary based on what the operators are actually doing and why provides a much needed diversification of their motivations for behavior – and allows revising assumptions about why they do what they do, anchoring these in what *is* rather than in what should be. This thesis shows that operators who are presumably established to function as autonomous entities not only have a hard time to escape the influence of local or national politics, but often need this political support to continue providing services. In one way or another, each operator sought validation and support within the political realm to continue to function. Allowing for more ways of discussing how water services are provisioned for, and in turn of the different types of support that operators may or may not receive, or the ability and interest or not to operate specific types of infrastructure, will provide much-needed inspiration for new ways of imagining and doing water services provisioning.

6. Concluding remarks

In sum, my argument is that taking operators and operations seriously is a promising entry-point for producing new insights and ideas about ways of doing and organizing the provision of water services. This taking seriously should happen from a place of empathy (if not sympathy), trying to understand what operators do and why by understanding their dilemmas, constraints, opportunities and motivations. The starting point should be that operators know what they are doing. Hence, rather than anchoring studies about water services provision in what should happen – in ideal-typical models – my plea is to anchor the production of new insights about how to organize the provision of water provisioning in what operators are actually doing, trying to understand this from their perspective. By documenting operator's practices in terms of what they are, rather than discussing them in terms of whether or not they comply with whatever norm or ideal is in vogue or required by donors or governments - either financial sustainability, universal access or a combination of the two where possible – a rich empirical basis can be provided to develop new language and new imaginations for organizing and learning about the provision of water services.

I base this plea importantly on the finding of this thesis that compliance – with a policy model – remains a mirage: operational realities are made to confirm the model. Even when operators do not comply, they are seldom defying the model. This combination of non-compliance and non-defiance takes a significant amount of energy and financial resources from all involved. Both from operators who need to reconcile degrees of compliance and defiance, and donors who are repeating the same type of project over and over again. I would like to advocate for a re-directing of this energy and the financial resources towards what I consider a more honest arrangement, one in which the donor community goes beyond technical assistance programs that try to 'correct' operators, forcing them to perform the more 'desired' practices associated with favorite policy models. Instead, I would like to see donors engaging in much more active efforts to learn with and from operators. I admit that this would require courage: the courage to acknowledge that much of what happens in the sector deviates from the policy dogma of commercialization and the courage to question the validity of the assumptions made by the model (including its problem-framing). It would require creating the political and social space to discuss, talk about and document the actual practices of operators, even if these sometimes defy or call into question cherished policy principles.

While this is a change that requires courage, I think that many of the actual modifications and adaptations or new solutions that emerge during implementation are, in fact, very pragmatic ones. I do not think that what is needed is necessarily a replacement of one governance model or system with another one. At the policy making level, the model of commercialization largely consists of (ideologically-informed) principles. I realize that changing these principles is a lengthy and fundamental process that would require 'conceptual learning', a change in paradigm. For this to happen, not only new principles need to be prioritized, but also relationships of cause and effect need to be re-imagined. The relationship between the introduction of private sector operators and efficiency[155], or community-based organizations and participation would need to be re-assessed, and incorporated in the talk. This type of policy learning (change) is time-consuming, and requires re-negotiations between all actors involved. I have shown that the convergence of interests across and dependencies between actors makes it unlikely that this will happen.

Yet, I think and have shown that relatively easier changes are possible and are in fact already happening. At an operational level, the adjustments and changes that operators engage in to make the model of commercialization work are relatively straightforward and consist of devising creative means to reach agreed outcomes or objectives. In the case of Mozambique, this involved the re-interpretation of the contracts with the private operator. In Uganda, deviations from the model consisted of more practical ways of channeling financial resources from the central to the local level, and in Indonesia this was visible in the way that local elites became engaged in water services provision to facilitate payment by consumers. I showed that these adjustments were relatively easy to implement, and happened through localized processes of trial and error – or indeed of learning. Such changes in for instance tariff setting, do not require buy-in from many other actors and can therefore be quickly done.

My plea is for such practical and pragmatic adaptations and modifications to be recognized and discussed across cases and contexts, to allow learning to happen. What is needed is the active creation and nurturing of spaces for such learning. Spaces where all involved are encouraged to discuss the details of what they do and use actual experiences as the basis for improvements, even when those doings or experiences deviate from the policy norm. In other words, financial viability, efficiency or the pricing of water can continue to be acknowledged as important, but there should be much more room to openly discuss subsidies as a very much needed source of

---

[155] Interesting in this respect is the shifting emphasis of the anti-privatization lobby. Rather than opposing the involvement of private sector organizations in water service provision, this lobby is increasingly focusing on the commercial principles of water service provisioning rather than the organizational form in which services are provided (McDonald, 2014).

support, or to acknowledge that political interference happens, without necessarily being detrimental to the provision of water services.

References

1. Adank, M. (2013). Small town water services: trends, challenges and models. Thematic Overview Paper 27. The Hague, The Netherlands: IRC International Water and Sanitation Centre.

2. ADB (2012). Indonesia: Country Water Assessment. Manila, Philippines: Asian Development Bank

3. Ahlers, R.; Cleaver, F.; Rusca, M. and Schwartz, K. (2014). Informal space in the waterscape: disaggregation and co-production. Water Alternatives, 7(1), 1-14.

4. Ahlers R., Schwartz K, Perez Guida V. The myth of 'healthy' competition in the water sector: The case of small scale water providers. Habitat International. Volume 38, April 2013, Pages 175-182.

5. Ahlers, R.; Perez Guida, V.; Rusca, M. and Schwartz, K. (2013). Unleashing entrepreneurs or controlling unruly providers? The formalization of small-scale water providers in Greater Maputo, Mozambique. The Journal of Development Studies, 49(4), 470-482.

6. Al'Afghani, M.; Paramita, D., Avessina, AV, Muhajir, MA and Heriati, F. (2015). The Role of Regulatory Frameworks in Ensuring The Sustainability of Community-Based Water and Sanitation. Center for Regulation Policy and Governance.

7. Anderson, J. (1975). Public-Policy Making. London, UK: Nelson.

8. Argyris, C. (1976). Theories of action that inhibit individual learning. American Psychologist, 31(9), 638-654.

9. Assies, W. (2003). David versus Goliath in Cochabamba: Water Rights, Neoliberalism, and the Revival of Social Protest in Bolivia. Latin American Perspectives, 30(3), 14-36.

10. Baietti, A.; Kingdom, W. Ginneken, M. (2006). Characteristics of well-performing public water utilities. Washington, DC.: World Bank

11. Bakker, K. (2003). Archipelagos and networks: urbanization and water privatization in the South. The Geographical Journal, 169(4), 328-341.

12. Bakker, K. (2007). The 'commons' versus the 'commodity': alter-globalization, anti-privatization and the human right to water in the global south. Antipode, 39(3), 430-455.

13. Bakker, K. (2010). Privatizing water: governance failure and the world's urban water crisis. Ithaca, New York: Cornell University Press.

14. Bakker, K. (2013a). Constructing 'public' water: the World Bank, urban water supply, and the biopolitics of development. Environment and Planning D: Society and Space, 31(2), 280-300.

15. Bakker, K. (2013b). 'Hegemony does not imply homogeneity: Thoughts on the marketization and privatization of water'. In: Harris, L. Goldin, J and Sneddon, C. (eds.), Contemporary Water Governance in the Global South: Scarcity, Marketization and Participation. New York: Routeledge.

16. BAPPENAS 2003. Strategi dan Rencana Aksi Keanekaragaman Hayati Indonesia 2003-2020 [Dokumen Nasional]. Jakarta, Indonesia: Badan Perencanaan Pembangunan Nasional (National Development Agency).

17. Bayliss, K. (2003). Utility Privatization in Sub-Saharan Africa: A case study of Water. The Journal of Modern African Studies, 41(4), 507-531.

18. Bel, G., Fageda, X. and Warner, M. (2010). Is private production of public services cheaper than public production? A meta-regression analysis of solid waste and water services, Journal of Policy Analysis and Management, 29(3), 553-577.

19. Bell, D. and Jayne, M. (2009). Small Cities? Towards a Research Agenda. International Journal of Urban and Regional Research, 33(3), 683-699.

20. Benson, D. and Jordan, A. (2011). What have we learned from policy transfer research? Dolowitz and Marsh revisited. Political Studies Review, 9(3), 366-378.

21. Berg, S. and Mugisha, S. (2010). Pro-poor water service strategies in developing countries: promoting justice in Uganda's urban project. Water Policy, 12(4), 589-601.

22. Blaustein, J. (2015). Speaking truth to power: policy ethnography and policy reform in Bosnia and Herzegovina. Oxford, UK: Oxford University Press.

23. Blokland, M.; Braadbaart, O. and Schwartz, K. (1999). Private Business, Public Owners: Government Shareholdings in Water Enterprises. The Hague, The Netherlands: Ministry of Housing, Spatial Planning and the Environment.

24. Boyne, G. (2002), 'Public and Private Management: What's the Difference?', Journal of Management Studies 39:1, Oxford, UK: Blackwell Publishers Ltd

25. Bozeman, B. (2004). All organizations are public: comparing public and private organizations. Washington, DC.: Beard Books.

26. BPPSPAM. (2015). Indonesia Water Utility Performance 2015. BPPSPAM (Drinking Water Supply System Development): Jakarta

27. Braadbaart, O.; Eybergen, N. and Hoffer, J. (2007). Managerial autonomy: does it matter for the performance of water utilities? Public Administration and Development, 27(2), 111-121.

28. Brocklehurst, C. Janssens, J. (2004), Innovative Contracts, Sound Relationships:Urban Water SectorReform in Senegal, Washington, D.C.: World Bank, Retrieved from: http://documents.worldbank.org/curated/en/610031468781529048/pdf/309470WSS0n o1011Senegal01public1.pdf.

29. Brown, A. (2002). Confusing Means and Ends: Framework of Restructuring, not Privatization, Matters Most. International Journal of Regulation and Governance, 1(2), 115-128.

30. Brunsson, N. (1989). The Organization of Hypocrisy: Talk, Decisions and Actions in Organizations. Chichester: Wiley. 242 pages

31. Brunsson, N. (1993). Ideas and actions: justification and hypocrisy as alternatives to control. Accounting, Organizations and Society, 18(6), 489-506.

32. Cairncross, S. (1992). Sanitation and water supply: practical lessons from the decade. Washington, DC.: World Bank.

33. Cairney, P. (2011). Understanding Public Policy: Theories and Issues. London, UK: Palgrave Macmillan.

34. Campbell, J. (2002). Ideas, Politics, and Public Policy. Annual Reviews of Sociology, 28, 21-38.

35. Carter, R.; Howsam, P. (1999). The Impact and Sustainability of Community water Supply and Sanitation Programmes in Developing Countries. Water and Environmental Journal, 13(4), 292-296.

36. Castro, J.E. (2007). Water governance in the twentieth-first century. Ambiente and Sociedade. 10(2), 97-118.

37. Castro, V. and Morel, A. (2008). Can delegated management help water utilities improve services to informal settlements?. Waterlines, 27(4), 289-306.

38. Christensen T. and Laegreid, P. (2002), 'A Transformative Perspective on Administrative Reforms', In: Christensen T. and Laegreid, P. (eds.), New Public management: The Transformation of Ideas and Practice, Hampshire: Ashgate Publishing Limited.

39. Choguill, C. (1989). Small Towns and Development: A Tale from Two Countries. Urban Studies, 26(2), 267-274.

40. Cleaver, F. (1999). Paradoxes of participation: questioning participatory approaches to development. Journal of International Development, 11(4), 597-612

41. Cleaver, F. and Whaley, L. (2018). Understanding process, power and meaning in adaptive governance: a critical institutional reading. Ecology and Society. 23 (2), Art. 49.

42. Cochrane, A. and Ward, K. (2012). Researching the geographies of policy mobility: confronting the methodological challenges. Environment and Planning A: Economy and Space, 44(1), 5-12.

43. Cohen, B. (2006). Urbanization in developing countries: current trends, future projections and key challenges for sustainability. Technology in Society, 28(1-2): 63-80.

44. Cohen, M.; March, J. and Olsen, J. (1972). A garbage can model of organizational choice. Administrative Science Quarterly, 17(1), 1-25.

45. Conca, K. (2006). Governing water contentious transnational politics and global institutions building. Global Environmental Accords. Cambridge: MIT Press.

46. Cornwall, A. and Gaventa, J. (2001). From users and choosers to makers and shapers: repositioning participation in social policy. IDS Working Paper 127. Sussex, UK: Institute of Development Studies.

47. deLeon, P. (1999). The missing link revisited: Contemporary implementation research. Review of Policy Research, 16(3-4), 311-338.

48. Denhardt, R. and Denhardt, J. (2000), The New Public Service: Serving Rather than Steering, Public Administration Review, 60(6), 549-559.

49. Dillery, R. (1945). Problems of the Small Town Water Works. Journal of American Water Works Association, 37 (11), 1185-1190.

50. DiMaggio, P. and Powell, W. (1983). The Iron Cage revisited: Institutional Isomorphism and Collective Rationality in Organizational Fields. American Sociological Review, 48(2), 147-160.

51. Doherty, B.; Haugh, H. and Lyon, F. (2014). Social Enterprises as Hybrid Organizations: A Review and Research Agenda. International Journal of Management Reviews, 16(4), 417-436.

52. Domberger S. and Jensen, P. (1997). Contracting out by the public sector: Theory, evidence, prospects. Oxford Review of Economic Policy, 13 (4), 67–78.

53. Easterby-Smith, M. (1997). Disciplines of Organizational Learning: contributions and critiques. Human Relations, 50(9), 1085-1113.

54. Economist, The (1994). 'Of Wets and Water', The Economist, March 26th 1994, London: Economist.

55. Falkernmark, M. (1977). UN Water Conference: Agreement on Goals and Action Plan. Ambio, 6(4), 222-227.

56. Foster, V. (1996). Policy Issues for the Water and Sanitation Sectors. Washington, DC: Inter-American Development Bank

57. Franceys, R. (2008). GATS, 'privatization' and institutional development for urban water provision: Future postponed'?. Progress in Development Studies, 8(1), 45-58.

58. Franceys, R. Cavill, S. and Trevett, A. (2016). Who really pays? A critical overview of the practicalities of funding universal access. Waterlines, 35(1), 78-93.

59. Fuente, D.; Gakii Gatua, J.; Ikiara, M.; Kabubo-Mariara, J. Mwaura, M. and Whittington, D. (2016). Water and sanitation service delivery pricing and the poor: An empirical estimate of subsidy incidence in Nairobi, Kenya. Water Resources Research, 52(6), 4845-4862.

60. Furlong, K. (2010). Neoliberal water management: trends, limitations, reformulations. Environment and Society: Advances in Research, 1(1), 46-75.

61. Furlong, K. (2015). Water and the entrepreneurial city: The territorial expansion of public utility companies from Colombia and the Netherlands. Geoforum, 58(2015), 195-207.

62. Gandy, M. (2010). Rethinking urban metabolism: water, space and the modern city. City, 8(3), 363-379.

63. Gerlach, E. and Franceys, R. (2010). Regulating Water Services for All in Developing Countries. World Development, 38(9), 1229-1240.

64. Ghosh, A. and Cairncross, S. (2014). The uneven progress of sanitation in India. Journal of Water, Sanitation and Hygiene for Development, 4(1), 15-22.

65. GOU-NWSC-PC5. (2015). Fifth Performance Contract for the period 1st July 2015 to 30th June 2018 between National Water and Sewerage Corporation and the Government of Uganda (GOU). Ministry of Water and Environment.

66. Graham, S. and Marvin, S. (2001). Splintering Urbanism: networked infrastructures, technologies mobilities and the urban condition. London, UK. Routeledge.

67. Haarmeyer, D. & Mody, A. (1997). Worldwide Water in Privatisation: Managing Risks in Water and Sanitation. London, UK: Financial Times Energy.

68. Hadipuro, W. (2010). Indonesia's water supply regulatory framework: Between commercialization and public service?. Water Alternatives, 3(3), 475-491.

69. Hall, D. (1999). Privatization, multinationals and corruption. Development in Practice, 9(5), 539-556.

70. Hall, D., Lobina, E. and Terhorst, P. (2013) Remunicipalisation in the early 21st century: Water in France and energy in Germany, International Review of Applied Economics, 27(2), 193-214.

71. Hall, P. (1993). Policy paradigms, social learning, and the State: the case of economic policymaking in Britain. Comparative Politics, 25(3), 275-296.

72. Harvey, P. and Reed, R. (2006). Community-managed water supplies in Africa: sustainable or dispensable?. Community Development Journal, 42(3), 365-378.

73. Heclo, H. (1974). Modern Social Politics in Britain and Sweden: from relief to income maintenance. New Haven, Connecticut. Yale University Press.

74. Hill, M. and Hupe, P. (2002). Implementing Public Policy: An Introduction to the Study of Operational Governance. Londong, UK: SAGE Publications, Inc.

75. Hopkins, R. and Satterthwaite, D. (2003). An Alternative Perspective on WSS services (including the 'grey areas'). In Appleton, B. (Ed). Town Water Supply and Sanitation Companion Papers: Volume 3. Bank Netherlands Water Partnership Project #043. Washington, DC.: The World Bank Group.

76. Hughes, O. (2003), Public Management and Administration: An Introduction, third edition, New York, US: Palgrave MacMillan.

77. Hughes, O. (2012). Public Management and Administration: An Introduction. London, UK: Red Globe Cross and MacMillan Higher Education

78. Hupe, P. and Sætren, H. (2015). Comparative Implementation Research: directions and dualities. Journal of Comparative Policy Analysis, 17(2), 93-102.

79. Hutchings, P.; Franceys, R.; Mekala, S.; Smit, S. and Jams, A. (2017). Revisiting the history, concepts and typologies of community management for rural drinking water supply in India. International Journal of Water Resources Development. 33(1): 152-169.

80. Hutton, G. and Varughese, M. (2016). The Costs of Meeting the 2030 Sustainable Development Goal Targets on Drinking Water, Sanitation, and Hygiene. Washington, DC.: World Bank. Retrieved from: https://openknowledge.worldbank.org/bitstream/handle/10986/23681/K8632.pdf?sequence=4 (last accessed 9 July 2019).

81. Idelovitch, E. and Ringskog, K. (1995). Private Sector Participation in Water Supply and Sanitation in Latin America. Washington D.C.: World Bank

82. Jager, U. and Beyes, T. (2010). Strategizing in NPOs: A case study on the practice of organizational change between social mission and economic rationale. Voluntas: International Journal of Voluntary and Nonprofit Organizations, 21(1), 82-100.

83. Jager, U. and Schroer, A. (2014). Integrated Organizational Identity: A definition of Hybrid Organizations and a Research Agenda. Voluntas: International Journal of Voluntary and Nonprofit Organizations, 25(5), 1281-1306.

84. Jaglin, S. (2008). Differentiating networked services in Cape Town: Echoes of splintering urbanism?. Geoforum, 39(6), 1897-1906.

85. Jensen, P. and Stonecash, R. (2005), Incentives and the Efficiency of Public Sector Outsourcing Contracts, Journal of Economic Surveys, 19(5), 767-787.

86. Kemp, R. and R. Weehuizen, S. (2005), Policy Learning, What Does It Mean and How Can We Study It?, Public Report No. D15, NIFU-STEP, Oslo.

87. Kippenberger, T. (1997). Some hidden costs of outsourcing. The Antidote, 2(6), 22-23.

88. Kitonsa, W. and Schwartz, K. (2012). Commercialization and centralization in the Ugandan and Zambian water sector. International Journal of Water, 6(3-4), 176-194.

89. Kleemeier, E.L., 2010. Rural private operators and rural water supplies: A desk review of experience. Water Unit, Sustainable Development Network. Washington D.C.: World Bank.

90. Komives, K. and Brook Cowen, P. (1999), Expanding Water and Sanitation Services to Low-Income Households: the Case of the La Paz-El Alto Conocession, Viewpoint, Note No. 178, Retrieved from: http://documents.worldbank.org/curated/en/421581468743380022/pdf/multi-page.pdf.

91. Komives, K. (2001). Designing pro-poor water and sewer concessions: early lessons from Bolivia. Water Policy, 3(1), 61-79.

92. Larner, W., & Laurie, N. (2010). Travelling technocrats, embodied knowledges: Globalising privatisation in telecoms and water. Geoforum, 41(2), 218-226.

93. Lauria, D.(2003). And Appropriate Design of Town Systems (A2). In Appleton, B. (Ed). Town Water Supply and Sanitation Companion Papers: Volume 3. Bank Netherlands Water Partnership Project #043. Washington, DC.: The World Bank Group.

94. Lee, T. and Floris, V. (2003). Universal access to water and sanitation: why the private sector must participate. Natural Resources Forum, 27(4), 279-290.

95. Lewis, MA.; and Miller TR. (1987). Public-private partnership in water supply and sanitation in Sub-Saharan Africa. Health Policy and Planning, 2(1), 70-79.

96. Lipsky, M. (1980). Street-level bureaucracy: dilemmas of the individual in public services. New York, US: Russell Sage Foundation.

97. Lobina, E. (2005) Problems with Private Water Concessions: A Review of Experiences and Analysis of Dynamics, International Journal of Water Resources Development, 21(1), 55-87.

98. Lockwood, H., & Le Gouais, A. (2011). Professionalising community-based management for rural water services (No. 2). Briefing Note. The Hague, The Netherlands: IRC.

99. Loftus, A. (2005). Free water as commodity: the paradoxes of Durban's water service transformations. In McDonald, D. and Ruiters, G. (eds.) The age of commodity: water privatization in South Africa. London, UK: Earthscan.

100. LP3ES. (2007). Kajian Cepat Terhadap Program-Program Pengentasan Kemiskinan Pemerintah Indonesia: Program WSLIC-2 dan PAMSIMASs. Jakarta, Indonesia. Water and Sanitation Program – The World Bank Group.

101. M.W.E Manual. (2013). Republic of Uganda Ministry of Water and Environment: Water Supply Design Manual Second Edition. Ministry of Water and Environment.

102. Mair, J.; Mayer, J. and Lutz, E. (2015). Navigating Institutional Plurality: Organizational Governance in Hybrid Organizations. Organization Studies, 36(6), 713-739.

103. Manor, J. (2004). User committees: A potentially damaging second wave of decentralisation? The European Journal of Development Research, 16(1), 192-213.

104. Mara, D. and Alabaster, G. (2008). A new paradigm for low-cost urban water supplies and sanitation in developing countries. Water Policy, 10(2008), 119-129.

105. Marin, P. (2009). Public-Private Partnerships for Urban Water Utilities: A Review of Experiences in Developing Countries. Washington, DC: World Bank:

106. Marsh, D. (1991). Privatization under Mrs. Thatcher: A Review of the Literature, Public Administration, 69(4), 459-480.

107. Marson, M. and Savin, I. (2015). Ensuring Sustainable Access to Drinking Water in Sub Saharan Africa: Conflict between Financial and Social objectives. World Development, 76(C), 26-39.

108. Matland, R. (1995). Synthesizing the implementation literature: the ambiguity-conflict model of policy implementation. Journal of Public Administration Research and Theory, 5(2), 145-174.

109. Maynard-Moddy, S.; Musheno, M. and Palumbo, D. (1990). Street-wise social policy: resolving the dilemma of street-level influence and successful implementation. Political Research Quarterly, 43(4).

110. Maynard-Moody, S. and Herbert, A. (1989).. Beyond implementation: Developing an institutional theory of administrative policy making. Public Administration Review, 49(2), 137-143.

111. Mbuvi, D. and Schwartz, K. (2013). The politics of utility reform: a case study of the Ugandan water sector. Public Money and Management, 33(5), 377-382.

112. McDonald, D. (2002). No money, no service. Alternatives Journal, 28(2), 16.

113. McDonald D. (2014). Public Ambiguity and the Multiple Meanings of Corporatization: Rethinking Corporatization and Public Services in the Global South. London: Zed Books.

114. McDonald, D. (2016). Making public in a privatized world: the struggle for essential services. London, UK: Zed Books.

115. McFarlane, K. and L. Harris. (2018). Small systems, big challenges: Review of small drinking water system governance. Environmental Reviews, 26(4), 378-395.

116. McInnes P. (2005). Entrenching Inequalities: The Impact of Corporatization on Water Injustices in Pretoria. In McDonald, D. A., & Ruiters, G. (2005). The age of commodity: Water privatization in Southern Africa. London, UK: Earthscan.

117. Mehta, L., Marshall, F., Movik, S., Stirling, A., Shah, E., Smith, A. and Thompson, J. (2007) Liquid Dynamics: challenges for sustainability in water and sanitation, STEPS Working Paper 6. Brighton, UK: STEPS Centre. Retrieved from: https://opendocs.ids.ac.uk/opendocs/handle/123456789/2464

118. Meyer, J. and Rowan, B. (1977). Institutionalized Organizations: Formal structures as Myth and Ceremony. American Journal of Sociology, 83(2), 340-363.

119. Meyers, M. and Vorsanger, S. (2005). Chapter 12: Street-level Bureaucrats and the Implementation of Public Policy. In Peters,G. and Pierre, J. (2005) (eds.). Handbook of Public Administration. Thousand Oaks, California: SAGE Publications.

120. Ministry of Water, Lands and Environment (1999). A National Water Policy. The Republic of Uganda.

121. Mitlin, D. (2008). With and Beyond the State – Co-production as a route to political influence, power and transformation for grassroots organizations. Environment and Urbanization, 20(2), 339-360.

122. Molle, F. (2008). Nirvana concepts, narratives and policy models: insights from the water sector. Water Alternatives, 1(1), 131-156.

123. Mollinga, P. (2008). Water, politics and development: Framing a political sociology of water resources management. Water Alternatives, 1(1), 7-23.

124. Moriarty, P., Patricot, G., Bastemeijer, T., Smet, J., & Van der Voorden, C. (2002). Between rural and urban: Towards sustainable management of water supply systems in small towns in Africa: Delft, Netherlands: IRC International Water and Sanitation Centre.

125. Moriarty, P., Smits, S., Butterworth, J. and Franceys, R. (2013). Trends in rural water supply: Towards a service delivery approach. Water Alternatives, 6(3), 329-349.

126. Mosse, D. (2004). Is good policy unimplementable? Reflections on the ethnography of aid policy and practice. Development and Change, 35(4), 639-671.

127. Moyson, S.; Scholten, P. and Weible, C. (2017). Policy learning and policy change: theorizing their relations from different perspectives. Policy and Society, 36(2), 161-177.

128. Mugabi, J. and Njiru, C. (2006). Managing water services in small towns: challenges and reform issues for low-income countries. Journal of Urban Planning and Development, 132(4), 187-192.

129. Muhairwe, W. (2009). Making Public Enterprises Work: From Despair to Promise: A Turn Around Account. London, UK. IWA Publishing Alliance House.

130. Mukhtarov, F. (2014). Rethinking the travel of ideas: policy translation in the water sector. Policy and Politics, 42(1), 71-88.

131. Mwanza, D. (2004). African Public Utilities Not Performing Efficiently. Paper presented at the 12th Union for African Water Suppliers Congress, 16 to19 February, 2004 in Accra, Ghana.

132. Mwanza, D. (2005). Promoting Good Governance through Regulatory Frameworks in African Water Utilities'. Water Science & Technology, 51(8), 71-79.

133. Nakamura, R. (1987). The Textbook Policy Process and Implementation Research. Policy Studies Review, 7(1), 142-154

134. Nalbandian, J. (2005). Professionals and the Conflicting Forces of Administrative Modernization and Civic Engagement, American Review of Public Administration, 35(4), 311-326.

135. Ndaw, M.F. (2016). Private Sector Provision of Water Supply and Sanitation Services in Rural Areas and Small Towns: The role of the Public Sector. Water and Sanitation Program: Guidance Note. Washington, DC: World Bank.

136. New Delhi Statement, Global Consultation on Safe Water and Sanitation, 1990. Report of the Economic and Social Council. Retrieved from: ielrc.org/content/e9005.pdf (accessed last 9 July 2019)

137. Nickson, A. (2002). The Role of the 'Non-State' Sector in Urban Water Supply. Paper for the 'Making Services Work for Poor People' World Development (WDR) 2003/04 Workshop, 4-5 November in Oxford. Retrieved from: http://siteresources.worldbank.org/INTWDR2004/Resources/22480_nicksonWDR.pdf

138. Njiru, C. (2004). Utility-small water enterprise partnerships: serving informal urban settlements in Africa. Water Policy, 6(5), 443-452.

139. Noronha de Vaz, T. (2014). Regional, national and international networks: the suitability of different competitive strategies for different geographic profiles. International Journal of entrepreneurship and Small Business, 21(3), 317-333.

140. NWSC SCAP100 (2016). 100% Water Service Coverage Acceleration project (SCAP100) in all villages under NWSC. NWSC. Kampala, Uganda.

141. NWSC. (2015). NWSC Infrastructure Service Delivery (ISDP) and Water Supply Stabilization Programme (WSSP): plan for the Financial Year 2015_2016. Retrieved from: https://www.nwsc.co.ug/index.php/contenthome/item/174-infrastructure-service-delivery-plans.

142. NWSC-Annual.Report. (2016). Intergrated Annual Report: Continuous Improvement for Sustainable and Equitable Service Delivery. Kampala, Uganda. Retrieved from NWSC-Corporate Strategy Department. Retrieved from: https://www.nwsc.co.ug/index.php/resources/reports

143. NWSC-CP (2018). NWSC Corporate Plan 2018-2021. Accelerated sustainable growth and service reliability through innovations. Kampala, Uganda. Retrieved from NWSC official website.

144. NWSC-SD. (2016). Five Year Strategic Direction 2016-2021. Kampala, Uganda. Retrieved from NWSC: Programmes and Performance Department

145. OECD (2009). Managing Water for All. An OECD Perspective on Pricing and Financing. Key Messages for Policy Makers. Paris: OECD.

146. OECD (2018). Financing water: investing in sustainable growth. Policy Perspectives: OECD Environment Policy Paper No. 11. Paris, France: OECD. Retrieved from: https://www.oecd.org/water/Policy-Paper-Financing-Water-Investing-in-Sustainable-Growth.pdf

147. Omuut, M. (2018). The impact of Infrastructure Development on Water Supply Services and Viability of Small Towns: National Water and Sewerage Corporation (NWSC) Bushenyi and Kitgum operational Areas. MSc Research Thesis. IHE Delft – Institute for Water Education. The Netherlands.

148. Panayotou, T. (1997). 'The Role of the Private Sector in Sustainable Infrastructure Development'. In Gomez-Echeverri, L. (ed.), Bridges to Sustainability: Business and Government Working Together for a Better Environment, Yale School of Forestry and Environmental Studies Bulletin Series 101. New Haven: Yale University.

149. Peck, J. and Theodore, N. (2010). Mobilizing Policy: models, methods and mutations. Geoforum, 41(2), 169-174.

150. Pigeon, M.; McDonald, D.; Hoedeman, O. and Kishimoto, S. (2012). Remunicipalization: Putting water back into public hands. Amsterdam, The Netherlands, Transnational Institute.

151. Pilgrim, N.; Roche, B.; Kalbermatten, J.; Revels, C. and Kariuki, M. (2007). Principles of Town Water Supply and Sanitation. Part 1: Water Supply. Water Working Notes No: 44223. Bank Netherlands Water Partnership. Washington, DC.: The World Bank Group.

152. Pitcher, A. (2002). Transforming Mozambique: the politics of privatization (1975-2000). Cambridge: Cambrigde University Press.

153. Pollit, C. (2001). Convergence: The useful myth?. Public Administration, 79(4), 933-947.

154. Pollit, C. and Hupe, P. (2011). Talking about government: the role of magic concepts. Public Management Review, 13(5), 641-658.

155. Polsby, N. (1969). Political science and the press: notes on coverage of a public opinion survey on the Vietnam war. Political Research Quarterly, 22(1), Art. 47.

156. Prasad, N. (2006). Privatisation results: private sector participation in water services after 15 years. Development Policy Review, 24(6), 669-692.

157. Price, S. and Franceys, R. (2003). Using private operators in small town water supplies, Uganda. Waterlines, 21(3), 19-21.

158. Raimundo, I. (2015). Urbanization of African cities: urban patterns in small-and medium-sized cities in Mozambique (unpublished).

159. Roa-Garcia, C. (2014). Equity, Efficiency and sustainability in water allocation in the Andes: trade-offs in a full world. Water Alternatives, 7(2), 298-319.

160. Roberts, B. H. (2016). Managing Systems of Secondary Cities: Policy Responses in International Development. Brussels, Belgium. Cities Alliance.

161. Robinson, J. (2005). Urban geography: world cities, or a world of cities. Progress in Human Geography, 29(6), 757-765.

162. Robinson, J. (2006). Ordinary cities: Between Modernity and Development. London: Routledge.

163. Rooyen, C. and Hall, D. (2007). Public is as private does: the confused case of Rand Water in South Africa. Kingston, Canada: Municipal Services Project.

164. Rothstein E. and Galardi, D. (2007). Financial sustainability as a foundation for infrastructure development and management: best practices Water Utility Management International, 2(1), 10-13.

165. Rondinelli, D. (1983). Towns and small cities in developing countries. Geographical review, 73(4), 379-395.

166. Rose, R. (1991). What is Lesson-Drawing? Journal of Public Policy, 11(1), 3-30.

167. Rusca, M. and Schwartz, K. (2018). The paradox of cost recovery in heterogeneous municipal water supply systems: ensuring inclusiveness or exacerbating inequalities? Habitat International, 73 (March): 101-108.

168. Rusca, M. Schwartz, K.; Hadzovic, L. and Ahlers, R. (2015). Adapting generic models through bricolage: elite capture of water user associations in peri-urban Lilongwe. The European Journal of Development Research, 27(5), 777-792.

169. Sabatier, P. and Jenkins-Smith, H. (eds) (1993). Policy Change and Learning: An Advocacy Coalition Approach. Boulder, Colorado: Westview Press.

170. Sabatier, P. (1986). Top-down and Bottom-up approaches to implementation research: a critical analysis and suggested synthesis. Journal of Public Policy, 6(1), 21-48.

171. Sabatier, PA. (2007). Fostering the development of policy theory. Theories of the policy process. Boulder, Colorado: WestviewPress.

172. Samanta, G. (2014). The politics of classification and the complexity of governance in census towns. Economic and Political weekly, 49(Issue No. 22).

173. Savelli, E.; Schwartz, K. and Ahlers, R. (2019). The Dutch aid and trade policy: policy discourses versus development practices in the Kenyan water and sanitation sector. Environment and Planning C: Politics and Space.

174. Schacter, M. (2000). Public Sector Reform in Developing Countries: Issues, Lessons and Future Directions. Prepared for Policy Branch Canadian international Development Agency. Institute on Governance. Ottawa, Canada: Institute on Governance.

175. Schaub-Jones D. (2008). Harnessing Entrepreneurship in the Water Sector: Expanding Water Services Through Independent Network Operators. Waterlines, 27(4).

176. Schiffler, M. 2015. Water, Politics and Money. Geneva, Switzerland: Springer International.

177. Schouten, M. and Buyi, T. (2010). International Journal of Public Sector Management, 23(5), 431-443.

178. Schwartz, K. (2006). Managing Public Water Utilities: An assessment of bureaucratic and New Public Management models in the water supply and sanitation sectors in low- and middle-income countries. PhD Dissertation. Erasmus University, Rotterdam, The Netherlands. Retrieved from: https://repub.eur.nl/pub/8052/

179. Schwartz, K. (2008). The New Public Management: The future of reforms in the African water supply and sanitation sector? Utilities Polity, 16(1), 49-58.

180. Schwartz, K. and Schouten, M. (2007). Water as a political good: revisiting the relationship between politics and service provision. Water Policy.,9(2), 119-129.

181. Scott, R. (2008). Institutions and Organizations: Ideas and Interests. Thousand Oaks, California. SAGE Publications, Inc.

182. Simon, H. (1982). Models of Bounded Rationality. Volume I: Economic analysis and Public policy. Cambridge, US: MIT Press.

183. Smith, L. (2003). The murky waters of the second wave of neoliberalism: corporatization as a service delivery model in Cape Town. Geoforum, 35(3), 375-393.

184. Spiller, P. and Savedoff, W (eds.) (1999). Spilled Water: Institutional Commitment in the Provision of Water Services. Washington, DC.: Inter-American Development Bank.

185. Solo, T.M.; Perez, E. and Joyce, S. (1993). Constraints in providing water and sanitation services to the urban poor. USAID WASH Technical Report No. 85.

186. Spronk, S. (2010). Water and Sanitation Utilities in the Global South: Re-Centering the Debate on "Efficiency". Review of Radical Political Economics, 42(2), 156-174.

187. Stone, D. (2012). Transfer and translation of policy. Policy Studies. 33(6): 483-499

188. Surya, R. (2017). Commercialization of Community Based Water Service Providers: the case of Lamongan Regency Water Service Provision. MSc Research Thesis. IHE Delft – Institute for Water Education. The Netherlands.

189. Swyngedouw, E (1995). Flows of Power: Water and the Political-Ecology of Urbanization". Human Geography Symposium in Honor of M. Gordon 'Reds' Wolman, The Johns Hopkins University, Baltimore, 1/01/24.

190. Tutusaus, M.; Cardoso, P. and Vonk, J. (2019). (de)Constructing the conditions for private sector involvement in small towns' water supply systems in Mozambique: policy implications. Water Policy, 20(S1), 36-51.

191. Tynan, N. and Kingdon, B. (2002). A Water Scorecard: setting performance targets for water utilities. Note Number 242. Washington DC.: The World Bank Group.

192. UNICEF (2017). Budget brief 2017: WASH. Mozambique. In collaboration with Forum de Monitoria do Orcamento and ROSC. Maputo, Mozambique: UNICEF.

193. UN-Water (2018). Nature-Based Solutions for Water: The United Nations World Water Development Report 2018. Report. Paris, France: UNESCO.

194. Wami, F. and Fisher, J. (2015). Effect of poor performance of water utilities in Port Harcourt city, Nigeria. Conference paper. WEDC 38th International Conference. Loughborough University. Retrieved from: https://dspace.lboro.ac.uk/dspace-jspui/handle/2134/31264

195. WaterAid (2013). From promise to reality: the urgent need for Southern African leaders to deliver on their water, sanitation and hygiene commitments. Briefing note. London, UK: WaterAid. Retrieved from: https://washmatters.wateraid.org/publications/from-promise-to-reality-urgent-need-for-southern-african-leaders-to-deliver-on-their

196. WaterAid (2014). Small town learning review: the synthesis report of a four country study. Report. London, UK: WaterAid. Retrieved from: https://washmatters.wateraid.org/.../small-town-water-and-sanitation-delivery-taking-a-...

197. WaterAid/BDP (2010). Small town water and sanitation delivery: taking a wider view. Report. London, UK: WaterAid. Retrieved from: http://www.bpdws.org/web/d/doc_265.pdf?statsHandlerDone=1

198. Whittington, D. (1992). Possible adverse effects of increasing block water tariffs in developing countries. Economic Development and Cultural Change, 41(1), 75-87.

199. WHO/UNICEF (2015). Joint Monitoring Program for Water Supply and Sanitation (JMP) – 2015 Update and MDG Assessment. WHO Press, World Health Organization – Geneva, Switzerland.

200. Winpenny J. (2015), Water: Fit to Finance?. Paris, France: World Water Council and OECD. Retrieved from: http://www.worldwatercouncil.org/sites/default/files/2017-10/WWC_OECD_Water-fit-to-finance_Report.pdf

201. Winpenny, J. (2003). Financing Water for All. Report of the World Panel on Financing Water Infrastructure: chaird by M. Camdessus. Marseille, France: World Water   Council: http://www.worldwatercouncil.org/fileadmin/world_water_council/documents_old/Library/Publications_and_reports/CamdessusReport.pdf (last accessed 9 July 2019)

202. Winter, S. (2012). Implementation Perspectives: Status and Reconsideration in in Peters,G. and Pierre, J. (2005) (eds.). Handbook of Public Administration. Thousand Oaks, California: SAGE Publications.

203. Wood, G. (1985). The Politics of Development Policy Labelling/ Development and Change, 16, 347-373.

204. World Bank (1994). World Development Report 1994: Infrastructure for Development. New York: Oxford University Press. Retrieved from: https://openknowledge.worldbank.org/handle/10986/5977

205. World Bank (1997). Best Practice in Urban Water Supply Côte d'Ivoire's SODECI - Capacity-building for Better Service, Retrieved from: http://documents.worldbank.org/curated/en/162081468244184867/pdf/597170BRI0Find10Box358292B01PUBLIC1.pdf.

206. World Bank (2017). Sustainability Assessment of Rural Water Service Delivery Models: Findings of a Multi-Country Review. Washington, DC.: World Bank. Retrieved from: https://openknowledge.worldbank.org/handle/10986/27988

207. World Health Organization & International Drinking Water Supply and Sanitation Decade (1981). Drinking-water and sanitation, 1981-1990: a way to health, a WHO contribution to the International Drinking Water Supply and Sanitation Decade. World Health Organization. Retrieved from: https://apps.who.int/iris/handle/10665/40127

208. World Health Organization. Community Water Supply and Sanitation Unit. (1992). The International Drinking Water Supply and Sanitation Decade: End of decade review (as at December 1990. World Health Organization. Retrieved from: https://apps.who.int/iris/handle/10665/61775

209. WSP (2002). Small Towns, Special Challenges. International Conference on Water Supply and Sanitation for Small Towns and Multi-Village Schemes. Addis-Ababa, Ethiopia 11-15 June 2002.

210. WSP (2011). The Hard Way to the High Road: transition of community-based Water Groups to Professional Service Providers in Indonesia. Water and Sanitation Program: Learning Note. Washington DC, US: World Bank.

211. WSP (2015). Water Supply and Sanitation in Indonesia: Turning Finance into Service for the Future. Service Delivery Assessment: May 2015. Washington, DC: The World Bank.

212. Yin, R. (2012). Applications of case study research. Thousand Oaks, California. SAGE Publications.

Appendices

Overview of interviews in Indonesia

No	Type of Organization	Organization	Location	Date	Name Interviewee	Position	Code
1	Central Government	National Supporting Agency for Water Supply System Development (BPPSPAM)	Jakarta	08/11/2016	Aulawi Dzin Nun	Board of BPPSPAM	CG1
2	Consultant	Individual Consultant	Jakarta	10/11/2016	Engkus Koesnadi	Finance Expert (involved In WB CBO project)	AC2
3	International	Water and Sanitation Program (World Bank)	Jakarta	09/01/2017	Deviriandy Setiawan	Finance and Institutions Expert	WB1
4	International	INDII (Indonesia Infrastructure Initative	Jakarta	09/11/2016	Popi Lestari	Impact Evaluation Adviser	INDII1
5	International	INDII (Indonesia Infrastructure Initative	Jakarta	09/11/2016	Timothy Ravis	Impact Evaluation Adviser	INDII2
6	Academia	Ibnu Khaldun Institute (associated to Dundee University)	Jakarta	12/01/2017	Dr. Mova Al Afghani	Regulatory Expert of CBOS	AC1
7	Province Government	Public Works Agency	Surabaya	14/11/2016	Shinta	Head of Water Supply Section	PG1
8	Province Government	Public Works Agency	Surabaya	14/11/2016	Heri Eko Purnomo	Head of Water Supply and Sanitation Division	PG2
9	Local Government	Public Works and Human Settlements Agency	Lamongan	15/11/2016	Agus Pindo	Head of Water Supply Section	LG1
10	Local Government	Planning and Development Agency	Lamongan	15/11/2016	Galih Yanuar	Heand of Human Settlements Section	LG2
11	CBO Association	HIPPAMS Banyu Urip Association	Lamongan	16/11/2016	Kiswanto	Trustee	BA1
12	CBO Association	HIPPAMS Banyu Urip Association	Lamongan	16/11/2016	Kasdan	Head of Association	BA2
13	CBO Association	HIPPAMS Banyu Urip Association	Lamongan	16/11/2016	Atekan	Head of Technical Divisions	BA3
14	Water Utility	PDAM Lamongan	Lamongan	24/11/2016	Ali	Head of Technical Divisions	PDM1
15	CBO1	HIPPAM Tlanak Village	Lamongan	17/11/2016	Panggeng Siswandi	Head of HIPPAM	HPM1A
16	CBO1	HIPPAM Tlanak Village	Lamongan	17/11/2016	Sri Rahayuningsih	Village's Head (ex-secretary HIPPAM)	VC1
17	CBO1	HIPPAM Tlanak Village	Lamongan	17/11/2016	Damis Indriyati	Treasury/Finance	HPM1C
18	CBO1	HIPPAM Tlanak Village	Lamongan	18/11/2016	Ika Setya	Finance Staff	HPM1D
19	CBO1	HIPPAM Tlanak Village	Lamongan	18/11/2016	Adbul Aziz	Technicall Staff	HPM1E
20	CBO1	HIPPAM Tlanak Village	Lamongan	17/11/2016	Sukirno	Consumer	HPMC1

No	Type of Organization	Organization	Location	Date	Name Interviewee	Position	Code
21	CBO8	HIPPAM Pengumbulanadi Village	Lamongan	21/11/2016	Zein	Head of HIPPAM	HPM8A
22		HIPPAM Pengumbulanadi Village	Lamongan	21/11/2016	Hadi	Treasury/Finance	HPM8B
23		HIPPAM Pengumbulanadi Village	Lamongan	22/11/2016	Eko	Consumer	HPM8C
24	CBO7	HIPPAM Pawer Siwa Village	Lamongan	25/11/2016	Mukhlasan	Head of HIPPAM	HPM7A
25		HIPPAM Pawer Siwa Village	Lamongan	25/11/2016	Mulyadi	Consumer	HPMC7
26		HIPPAM Kemlagi Gede Village	Lamongan	28/11/2016	Munaji	Head of HIPPAM	HPM2A
27	CBO2	HIPPAM Kemlagi Gede Village	Lamongan	28/11/2016	Suyitno	Village's Head	VC2
28		HIPPAM Kemlagi Gede Village	Lamongan	28/11/2016	Nurkhozin	Consumer	HPMC2
29		HIPPAM Trepan Village	Lamongan	01/12/2016	Sapari	Head of HIPPAM	HPM6A
30	CBO6	HIPPAM Trepan Village	Lamongan	01/12/2016	Djuwoto	Village's Head	HPM6B
31		HIPPAM Trepan Village	Lamongan	05/12/2016	Panjo	Technical Staff	HPM6C
32		HIPPAM Trepan Village	Lamongan	06/12/2016	Kasmanto	Consumer	HPMC6
33		HIPPAM Geger Village	Lamongan	06/12/2016	Mudofar	Treasury/Finance	HPM3A
34	CBO3	HIPPAM Geger Village	Lamongan	08/12/2016	Suid	Secretary	HPM3B
35		HIPPAM Geger Village	Lamongan	07/12/2016	Widya	Consumer	HPMC3
36		HIPPAM Sukomulyo Urban Communities (Kelurahan)	Lamongan	12/12/2016	Yoyok Tri	Head of HIPPAM	HPM5A
37		HIPPAM Sukomulyo Urban Communities (Kelurahan)	Lamongan	12/12/2016	Askuri Anam	Treasury/Finance	HPM5B
38	CBO5	HIPPAM Sukomulyo Urban Communities (Kelurahan)	Lamongan	14/12/2016	Sigit W	Pembaca Meter/Technical Staff	HPM5C
39		HIPPAM Sukomulyo Urban Communities (Kelurahan)	Lamongan	14/12/2016	Kusman	Consumer	HPMC5
40		HIPPAM Karangwedero Village	Lamongan	18/12/2016	Witono	Head of HIPPAM	HPM4A
41	CBO4	HIPPAM Karangwedero Village	Lamongan	19/12/2016	Slamet	Secretary/Technical Staff	HPMA4B
42		HIPPAM Karangwedero Village	Lamongan	18/12/2016	Amin	Consumer	HPMC4
43	Project NUFFIC 196	Workshop organized by Minitry of Public Works (BTAMII) for CBOs from 9-13 October 2017. M. Tutusaus lead 50% of the facilitation					

Overview of interviews in Uganda

N	Station	Location		Date	Name	Position	Code
1	NWSC Hqtr	Head Office		15/11/2017	Eng. Alex Gisagara	Director Engineering services	Hq1
2	NWSC Hqtr	Head Office		16/11/2017	Eng. Lawrence Muhirwe	Sn. Manager Operations N&E region	Hq2
3	NWSC Hqtr	Head Office		16/11/2017	Prof. Mahmood Lutaya	Sn. Manager Operations S&W Region	Hq3
4	NWSC Hqtr	Head Office		13/01/2018	Mr. Silver Emudong	Sn. Manager Finance and Accounts	Hq4
5	NWSC Hqtr	Head Office		10/01/2018	Mr. Jude Mwoga	Sn. Manager Programmes and Performance Monitoring	Hq5
6	NWSC Hqtr	Head Office		22/11/2017	Eng. Denis Taremwa	Manager Water supply Infrastructure Dev't	Hq6
7	NWSC Hqtr	Head Office		24/11/2017	Eng. Cyrus Aomu	Principle Eng. Planning and Capital Dev't	Hq7
8	NWSC Hqtr	Head Office		05/12/2017	Aaron Magara	Regional Engineer N&W region	Hq8
9	NWSC Hqtr	Head Office		05/12/2017	Geoffrey Dwoka	Regional Engineer N&W region	Hq9
10	NWSC Hqtr	Kampala		15/05/2017	Edmond Okoron	Regional Manager Eastern Uganda	Hq9
	Ministry of Water and Environment	Kampala		15/05/2017		Representative of Ministry and Umbrella organizations	NG1
11	Bushenyi/Ishaka	Bushenyi/Ishaka Area	Operational	07/12/2017	Francis Kateeba	Area Manager	BI1
12	Bushenyi/Ishaka	Bushenyi/Ishaka Area	Operational	07/12/2017	Rogers Mugabe	Area Engineer	BI2
13	Bushenyi/Ishaka	Bushenyi/Ishaka Area	Operational	07/12/2017	Francis Oluka	Area accounts Officers	BI3
14	Bushenyi/Ishaka	Bushenyi/Ishaka Area	Operational	08/12/2017	Owona John Bosco	Branch Manager Ishaka	BI4
15	Bushenyi/Ishaka	Bushenyi/Ishaka Area	Operational	08/12/2017	Alex Ashabahebwa	Commercial Officer Billing/Revenue	BI5
16	Bushenyi/Ishaka	Bushenyi/Ishaka Area	Operational	09/12/2017	Peter Engwanyu	shift overseer	BI6
17	Bushenyi/Ishaka	Bushenyi/Ishaka Area	Operational	09/12/2017	Mr.John	Plumber	BI7
18	Bushenyi/Ishaka	Bushenyi/Ishaka Area	Operational	17/05/2017		Branch Manager Kabowhe	BI8
19	Bushenyi/Ishaka	Bushenyi/Ishaka Area	Operational	18/05/2017	Charles Mushabe	Branch Manager Rubirizi	BI9
20	Private operator	Bushenyi/Ishaka Area	Operational	17/05/2017	Damson Nuyambi	Ex-manager Kabowhe Branch (private operator)	BI10

N	Station	Location	Date	Name	Position	Code
21	Private operator	Bushenyi/Ishaka Operational Area	18/05/2017	David Akena	Ex-manager Rubirzi Branch (private operator)	BI11
22	Bushenyi/Ishaka	Bushenyi/Ishaka Operational Area	17/05/2017	Eng Israel	Former Area Engineer	BI12
23	Local Government	Bushenyi/Ishaka Operational Area	16/05/2017	Katunda Mutura	Town clerk (later accompanied by Mayor Bushenyi/Ishaka)	LG1
24	Local Government	Bushenyi/Ishaka Operational Area	16/05/2017		Standpipe operator Bushenyi/Ishaka main road	BI13
25	Kitgum Area	Kitgum Operational Area	13/12/2017	Faith Nambuya	Area Manager	KA1
26	Kitgum Area	Kitgum Operational Area	13/12/2017	Samson Munanura	Area Engineer	KA2
27	Kitgum Area	Kitgum Operational Area	15/12/2017	Patrick Opio	Area Accounts Officer	KA3
28	Kitgum Area	Kitgum Operational Area	15/12/2017	Fred Bongomin	Commercial Officer Billing/Revenue	KA4
29	Kitgum Area	Kitgum Operational Area	14/12/2017	Moses Okello	Plant Operator	KA5
30	Kitgum Area	Kitgum Operational Area	16/12/2017	Owona B	Plumber	KA6
31	Project SMALL	Head Office NWSC	19-22 February 2018	Workshop with Branch/Area Managers of Northern and Eastern Region NWSC. Facilitated by M.Tutusaus		WK1

Overview interviews in Mozambique

No	Type of Organization	Organization	Location	Date	Name Interviewee	Position	Code
1	Local Government	AIAS	Maputo	11/06/2015	Arlindo Fernando	Head of Engineering Department	AIAS9
2	National Government	AIAS	Maputo	10/06/2015	Elcina Mulambo	Director Commercial Relations	AIAS3
3-4	National Government	AIAS	Maputo	26/06/2015 and 12/02/2016	Valdemiro Matavela	Head of Technical Department	AIAS2
5-7	National Government	AIAS	Maputo	24/11/2015; 14/02/2016 and 05/02/2018	Frederico Martins	Strategic Advisor to AIAS (former FIPAG Director)	AIAS1
8	National Government	AIAS	Maputo	05/02/2018	Pedro Manjate	Legal Advisor	AIAS7
9	National Government	AIAS	Maputo	25/11/2015	Enginhero Rocha	Junior Engineer	AIAS8
10	National Government	AIAS	Maputo	07/02/2018	Rute Nhamucho	Current Executive Director	AIAS4
11	National Government	AIAS	Maputo	07/02/2018	Oulinda de Sousa	Former Executive Director	AIAS5
12	National Government	AIAS	Maputo	07/02/2018	Pedrito Antonio	Former (first) Executive Director	AIAS6
13	National Government	FIPAG	Chokwe	23/02/2016	Jose Chiure	Director Chokwe Branch	FIPAG1
14	National Government	CRA	Maputo	26/06/2015	Antonio Mirasse	Head Compliance	CRA6
15	National Government	CRA	Maputo	25/11/2015	Manuel Alvarinho	Former President	CRA1
16	National Government	CRA	Maputo	26/11/2015	Clara Dimene	Head of Legal	CRA3
17	National Government	CRA	Maputo	06/02/2018	Dinis Juizo	Advisory Board member	CRA4
18	National Government	CRA	Maputo	16/02/2016	Jordi Gallego	Consultant to CRA	CRA5
19	National Government	CRA	Maputo	09/02/2018	Suzana Saranga	Current President and former Head of DNA	CRA2
20	National Government	DNASS	Maputo	12/06/2015	Felicidade Masingue	Policy Adviser	DNASS1
No	Type of Organization	Organization	Location	Date	Name Interviewee	Position	Code
21				14/02/2016			
22	National Government	DNASS	Maputo	24/11/2015	Arlindo Correia	Senior Policy Advisor	DNASS2

#	Category	DNASS	City	Date	Name	Director WASH Division	DNASS3
23	National Government	DNASS	Maputo	19/02/2016	Nilton Trinidade	Director WASH Division	DNASS3
24	Private sector	Aguas de Caia	Caia	15/06/2015	Arlindo Bhikha	Director (Private operator)	PO1
25	Private sector	A Gota (water system)	Maputo	26/11/2015	Francisco Guambe	Director (Private operator)	PO2
26	Private sector	Collins Ltd	Maputo	26/11/2015	Pedro Cardoso	Director (Private operator)	PO3
27	Private sector	Aguas Manjacaze	Manjacaze	24/02/2016	Abtecio Marubini	Technical Staff (private operator)	PO4
28	Private sector	Aguas do Bilene	Bilene	22/02/2016	Elicio Ashmal	Technical Staff (private operator)	PO5
29	Private sector	Collins Ltd	Maputo	18/02/2016	Taurai Tomse	Technical Staff (private operator)	PO6
30	International	UNICEF	Maputo	23/06/2015	Alfonso Alverestegui	Head WASH Unit	PO6
31	International	UNICEF	Maputo	05/02/2018	Jesus Tralles	WASH Unit	INT2
32	International	VEI	Maputo	22/06/2015	Joep Vonk	Project Manager	INT3
33				23/11/2015			
34	International	WSP (World Bank)	Maputo	12/06/2015	Pedro Simone	Project Officer	INT4
35				19/02/2016			
36	International	WSP (World Bank)	Maputo	27/11/2015	Luis Macario	Senior Project Officer	INT5
37	International	EU Commission Representation	Maputo	23/06/2015	Thierry Rivol	Water Section EU Commission	INT6
38	Local Government	District Infrastructure Office	Caia	18/06/2015	Celso Jose Vasco	Head Engineer	LG1
39	Local Government	Municipality	Namaacha	24/06/2015	Gustavo Zita	Technical Advisor	LG2
40	Local Government	District Infrastructure Office	Moamba	18/02/2016	Isaura Manuel	Head Engineer	LG3
41	Local Government	District Government Caia	Caia	17/06/2015	Albino Chimundo	Governor	LG4
42	Local Government	Municipality	Moamba	18/02/2016	Jorge Madlava	Village's Head	LG5
43	Local Government	Municipality	Chokwe	23/02/2016	Several people	Water Department	LG6
44	Local Government	Municipality	Bilene	22/02/2016	Several people	Water Department	LG7
45	Local Government	Municipality	Manjacaze	24/02/2016	Several people	Water Department	LG8
46	Academia	Universidade Eduardo Mondlane	Maputo	22/06/2015	Ines Raimundo	Director Center of Political Analysis	AC1
47	Note	Collins Ltd and VEI have been partners in project SMALL. I have been in regular informal contact with representatives of these organizations until the finalization of this thesis. Some information was gained during these interactions.					

About the Author

Mireia Tutusaus Luque studied international business administration at the Universitat Pompeu Fabra in Barcelona (Spain) and holds a Master's degree in Water Services Management from IHE Delft Institute for Water Education (formerly known as UNESCO-IHE) in the Netherlands. She joined IHE Delft in 2014 where she worked on her PhD thesis while filling academic posts as lecturer, researcher, consultant and program coordinator of the Water Management and Governance MSc program at IHE Delft. Prior to joining IHE Delft she worked over six years in the private sector, first in retail banking in Spain and later on for an international corporation in the Netherlands where she fulfilled several positions from customer care and logistics to short-term strategic planning and production.

Her first encounters with research where at IHE Delft during the development of her MSc thesis. Mireia studied the development and positioning of informal water providers in the outskirts of Maputo, in Mozambique. Undertaking this research led to an interest in the daily works of small-scale water providers. This topic, later on, developed into the topic of her PhD. Her main area of expertise and interest in the field of water services management has been the development and analysis of business and governance models of the provision of drinking water services in urban areas and small towns, with a specific interest on financial sustainability and financial implications of infrastructural development. She has studied different forms of water service provisioning in different countries such as Indonesia, Uganda, Kenya, Mozambique and Tanzania. She has also collaborated in research projects focusing on the adoption of innovative technologies for water services in Europe.

At IHE Delft Mireia teaches in various topics such as policy analysis and organizational management in different modules such as Managing Water Organizations, Finance in the Water Sector and Partnerships for Water and Sanitation. She also supervises MSc students in their thesis studies in related topics.

Mireia currently resides in Kigali, Rwanda, working as a resident expert for an assignment with VEI Dutch Water Operators (former Vitens-Evides International), where she tries to combine, to the best of her capabilities, her academic background and experience gained in utility management at IHE Delft with her decisiveness to deliver fair and equitable access to water.

List of publications related directly to this thesis

Tutusaus, M.; Schwartz, K (eds.) (2018). Water services in small towns in developing countries: At the tail of development. Water Policy. 20: 1-11

Mireia Tutusaus and Klaas Schwartz contributed to the design and implementation of this Special Issue. The analysis of the results and the writing of the manuscript was carried out equally by both authors.

Tutusaus, M. Cardoso, P. and Vonk, J. (2018). (De)Constructing the conditions for private sector involvement in small towns' water supply systems in Mozambique: Policy implications. Water Policy. 20: 36-51.

Mireia Tutusaus designed and directed the research project. Mireia Tutusaus carried out the date collection, elaborated on the analysis of the results and carried out most of the writing. Pedro Cardoso and Joep Vonk provided additional data relevant to finalize the paper and commented on the manuscript.

Tutusaus, M.; Schwartz, K. and Surya, R. (under revision). Degrees and forms of commercialization: community-managed water operators in Lamongan Regency, Indonesia. Water,11, 1985.

Mireia Tutusaus designed and directed the project, contributed to the analysis of the data and conceived of the presented ideas. Klaas Schwartz supervised the work and contributed to the theoretical framework. Mireia Tutusaus and Klaas Schwartz contributed equally to the writing of the manuscript. Maxi Omuut collected raw data and participated in the analysis of the data

Tutusaus, M.; Schwartz, K. and Omuut, M. (submitted). Commercialization as Organized Hypocrisy: The Divergence of Talk and Actions in Water Services in Small Towns in Uganda. Water Alternatives

Mireia Tutusaus designed and directed the project, contributed to the analysis of the data and conceived of the presented ideas. Klaas Schwartz supervised the work and contributed to the theoretical framework. Mireia Tutusaus and Klaas Schwartz contributed equally to the writing of the manuscript. Maxi Omuut collected raw data and participated in the analysis of the data.

Schwartz, K. and **Tutusaus, M.** (forthcoming). Legal Frameworks and Water Services: A case of Confused Identities. In Dellapenna, J. and Gupta, J. (eds). Elgar Encyclopedia on Environmental Law. Volume: Water Law. Edward Elgar Publishing.

Klaas Schwartz and Mireia Tutusaus contributed to the conceptualization and writing of this book chapter in equal parts.

Other peer-reviewed articles:

- Kemendi, T. and **Tutusaus, M.** (2018). 'The impact of pro-poor service units on the performance of utilities: the case of Nakuru and Kisumu'. Journal of Water, Sanitation and Hygiene for Development.
- **Tutusaus, M.**, Schwartz, K. (2018). Water services in small towns in developing countries: At the tail of development. Water Policy. 20: 1-11.
- Schwartz, K.; Gupta, J. and **Tutusaus, M.** (2018). Editorial-Inclusive development and urban water services. Habitat International. 73: 96-100.
- **Tutusaus, M.**, Schwartz, K. and Smit, S. (2018). The ambiguity of innovation drivers: The adoption of information and communication technologies by public water utilities. Journal of Cleaner Production. 171: S79-S85.
- Zwarteveen, M., J.S. Kemerink-Seyoum, M.Kooy, J. Evers, T.A.Guerrero, B.Batubara, A.Biza, A.Boakye-Ansah, S. Faber, A.C. Flamini, G Cuadrado-Quesada, E.Fantini, J. Gupta, S.Hasan, R. ter Horst, H.Jamali, F.Jaspers, P. Obani, K.Schwartz, Z. Shubber, H. Smit, P. Torio, **M.Tutusaus** and Anna Wesselink (2017). Engaging with the politics of water governance, Opinion piece, WIREs Water, 2017, e01245. doi: 10.1002/wat2.1245.
- Schwartz, K., **Tutusaus, M.** and Savelli, E. (2017), 'Water for the Poor Water for the Urban Poor: Balancing financial and social objectives through service differentiation for low-income areas in the Kenyan water services sector. Utilities Policy. 48: 22-31.
- **Tutusaus, M.**, Schwartz, K. and Smit, S. (2016), 'The Ambiguity of Innovation Drivers: The Adoption of Information and Communication Technologies by Public Water Utilities', *Journal of Cleaner Production.*
- **Tutusaus, M.** and Schwartz, K. (2016), 'National Water Operators' Partnerships: a promising instrument for capacity development?', *Journal of Water, Sanitation and Hygiene for Development*, 6(3), pp. 500-506.
- Schwartz, K., **Tutusaus, M.**, Rusca, M. and Ahlers, R. (2015), '(In)formality: The Meshwork of Water Service Provisioning', *WIREs Water*, 2(1):31–36.

Special Issues (guest editor)

Tutusaus, M.; Schwartz, K (eds.) (2018). Water services in small towns in developing countries: At the tail of development. Water Policy. 20: 1-11

Schwartz, K.; Gupta, J. and **Tutusaus, M**. (2018). Inclusive development and urban water services. Habitat International. 73: 96-100 *(participated in the editorial chapter)*